Wireless Networks

Series Editor

Xuemin Sherman Shen, University of Waterloo, Waterloo, Canada

The purpose of Springer's Wireless Networks book series is to establish the state of the art and set the course for future research and development in wireless communication networks. The scope of this series includes not only all aspects of wireless networks (including cellular networks, WiFi, sensor networks, and vehicular networks), but related areas such as cloud computing and big data. The series serves as a central source of references for wireless networks research and development. It aims to publish thorough and cohesive overviews on specific topics in wireless networks, as well as works that are larger in scope than survey articles and that contain more detailed background information. The series also provides coverage of advanced and timely topics worthy of monographs, contributed volumes, textbooks and handbooks.

Rong Zheng • Chao Cai

Acoustic Sensing on Commodity Devices and its Applications

Springer

Rong Zheng
Department of Computing and Software
McMaster University
Hamilton, ON, Canada

Chao Cai
College of Life Science and Technology
Huazhong University of Science and Technology
Wuhan, China

ISSN 2366-1186 ISSN 2366-1445 (electronic)
Wireless Networks
ISBN 978-3-031-96874-7 ISBN 978-3-031-96875-4 (eBook)
https://doi.org/10.1007/978-3-031-96875-4

This Springer imprint is published by the registered company Springer Nature Switzerland AG
The registered company address is: Gewerbestrasse 11, 6330 Cham, Switzerland

To my mom, whose boundless curiosity continues to inspire me.

—Rong

To my family, whose endless love supports me.

—Chao

Preface

Acoustic transducers, such as speakers and microphones, are widely available in everyday devices such as smartphones, home entertainment systems, smart doorbell, and vehicle infotainment systems. Beyond their primary functions of sound recording and playback, these transducers can be repurposed to sense our surroundings and human activities. This capability has unlocked a plethora of opportunities for both the research community and industry, leading to novel applications such as proximity and presence detection of humans, localization and tracking of smart devices, non-contact vital sign estimation, and mid-ear infection detection. These innovations demonstrate the untapped potential of acoustic sensing.

The primary advantage of leveraging acoustic sensing on existing commodity devices is its accessibility and cost-effectiveness compared to deploying specialized hardware solutions. However, designing such systems requires overcoming challenges posed by hardware constraints and the limitations of the human auditory system. For instance, commodity devices are often not designed with high-precision sensing in mind, necessitating customized waveform design and signal processing to achieve robust performance. Despite these challenges, the field of acoustic sensing has made significant strides in the past decades.

This book focuses on the techniques and applications of **active acoustic sensing**, which uses an acoustic source, such as a speaker, to generate sound waves. These waves travel through air or solid medium and interact with objects and human in the environment. By analyzing the reflected or directly received acoustic signals, it is possible to infer valuable information about the surroundings, enabling diverse sensing applications. Despite the growing interest in active sensing, there has not been a comprehensive treatment on the core techniques and new advancements in literature. The book provides an in-depth coverage of the basic building blocks and algorithms for active acoustic sensing on commodity devices, design considerations, and novel applications they enable.

The book is intended for a diverse audience. Researchers and engineers interested in developing active acoustic sensing solutions on commodity devices will find a detailed account of state-of-the-art techniques and results in the field. Graduate students looking to enter this exciting domain will benefit from the book's coverage of foundational knowledge, accompanied by practical implementations of key algorithms in Matlab. These implementations aim to bridge the gap between theory and practice, providing readers with a hands-on understanding of the concepts discussed.

Hamilton, ON, Canada Rong Zheng
Wuhan, Hubei, China Chao Cai
February, 2025

Acknowledgments

We would like to thank our families and funding agencies for the continuing support of the work. Some of the research work included is not possible without the help of our students, Yuting Wang, Jun Li, Navid Hossein Zandi, Bodee Quansah, Henglin Pu, Peng Wang, Liyuan Ye, Pei Rao, Yan Yan, and Meichen An, and collaborators Dr. Dongmei Zhao, Dr. Ian Bruce, and Dr. Jun Luo.

Contents

Acronyms

AAC	Aerial acoustic communication
AGC	Automatic gain control
AoA	Angle of arrival
AR	Augmented reality
BER	Bit error rate
BPF	Band-pass filter
CFO	Carrier-frequency offset
CIR	Channel impulse response
CSI	Channel state information
FFT	Fast Fourier transform
FMCW	Frequency-modulated continuous wave
FSK	Frequency shift keying
GPS	Global positioning system
HCI	Human-computer interaction
HRIR	Head-related impulse response
HRTF	Head-related transfer function
IFFT	Inverse fast Fourier transform
IMU	Inertial measurement unit
IoT	Internet of Things
ISI	Inter-symbol interference
LOS	Line-of-sight
LPF	Low-pass filter
MIMO	Multiple-input multiple-output
NLOS	None-line-of-sight
OFDM	Orthogonal frequency division multiplexing
PAPR	Peak-to-average-power-ratio
RMS	Root mean square
SFO	sampling frequency offset
SNR	Signal-to-noise ratio
SPL	Sound pressure level
SS	Spectrum spread

TDMA	Time division multiple access
TDoA	Time difference of arrival
ToA	Time of arrival
ToF	Time of flight
ULA	Uniform linear array
VR	Virtual reality

Chapter 1
Introduction

1.1 Overview of Acoustic Sensing

Internet of Things (IoT) technologies enable everyday objects to connect and communicate with each other by augmenting them with sensing, processing, and computation capability [34]. With the ever -increasing computation power and rich built-in sensors available in IoT devices, novel applications emerge by repurposing existing sensors beyond their primary use. For instance, cameras, originally designed for capturing photos, have been repurposed for use in visible light communication [35]. Gyroscope and accelerometer sensors are designed for pose and motion estimation but have been used extensively in human activity recognition [15]. WiFi signals, widely used for facilitating communication, have also been applied in many context-aware computing applications including localization [1, 12, 29] and gesture recognition [22, 30]. In this monograph, we target innovative sensing techniques and applications that exploit acoustic front ends of commodity IoT and mobile devices.

Acoustic front ends, namely microphones and speakers, are among of the most commonly available transducers in IoT devices. They are primarily designed for playing back and recording audio signals. However, they can also be repurposed for novel applications in acoustic sensing. A sensor is a device that produces an output signal given a certain physical stimuli, while sensing is the process of producing such a signal. *Acoustic sensing* is thus the process of using acoustic transducers to produce electric signals from vibration in the medium. *Passive acoustic sensing* involves detecting and analyzing sound waves in the environment without actively emitting any signals. Instead of generating acoustic signals, passive acoustic sensors capture existing sounds or vibrations, such as those produced by human activities, machinery, or natural phenomena. Doing so enables monitoring and interpreting various acoustic events, often for applications in security, wildlife monitoring, industrial diagnostics, and environmental studies. By relying solely on ambient noise, passive acoustic sensing provides a nonintrusive and energy-efficient way

R. Zheng, C. Cai, *Acoustic Sensing on Commodity Devices and its Applications*, Wireless Networks, https://doi.org/10.1007/978-3-031-96875-4_1

to gather information about the surroundings. *Active acoustic sensing*, in contrast, involves emitting acoustic signals into the environment and then detecting and analyzing the received sound waves. This method uses an acoustic source, such as a speaker or transducer, to generate sound waves, which travel through the medium (air, water, or solid materials) and interact with objects in the environments. Acoustic sensors capture signals propagated along the direct path or via reflections, and the data is processed to infer information about the objects and their characteristics, such as distance, shape, size, and movement. Active acoustic sensing has been applied in applications such as sonar, medical imaging (ultrasound), structural health monitoring, and object detection and ranging. This mechanism provides precise and controlled measurement capabilities through purposefully designed acoustic waveforms, making it useful for a variety of scientific, industrial, and security purposes. Traditionally, *acoustic communication* refers to the transmission and reception of information via sound waves. It shares significant similarities with active acoustic sensing, particularly in the use of customized (modulated) waveforms and the processing of received signals to extract meaningful information. Despite differing design objectives—acoustic communication focuses on transmitting source information, while active acoustic sensing emphasizes estimating channel information—we adopt a broader perspective, treating acoustic communication as a specialized form of active acoustic sensing. Such a treatment is further justified by the growing interests in integrated communication and sensing [16].

The primary focus of this book is *active acoustic sensing on acoustic-enabled commodity devices*. For simplicity, unless noted otherwise, we will refer to active acoustic sensing as acoustic sensing throughout the text. *Commodity devices* refer to widely available, mass-produced hardware products designed for general consumer use. Examples of acoustic-enabled commodity devices include smartphones equipped with microphones and speakers and IoT devices with embedded acoustic transducers. They offer a cost-effective and scalable platform for deploying active acoustic sensing applications. These devices' ubiquity and accessibility reduce the barriers to adoption. However, enabling acoustic sensing on commodity devices is not without its challenges. Specifically:

- **Device heterogeneity:** Unlike dedicated acoustic devices, the acoustic front ends on commodity devices are inherently limited in several aspects, including their number, frequency response, placement, and orientation. These constraints stem from the primary design goals of commodity devices, which prioritize general-purpose functionality, cost efficiency, and compact form factors over specialized acoustic performance. Furthermore, even among devices of the same type—such as smartphones—these characteristics can vary significantly between manufacturers, models, and even production batches. This variability introduces substantial challenges for researchers, as solutions designed for one configuration may not perform well across other devices. To address this, researchers must develop robust, adaptable solutions that account for the diversity and limitations of commodity hardware.

- **Deployment heterogeneity:** The deployment scenarios of these devices, along with the diversity of human interactions, present a wide range of variations, making system design particularly challenging. Environmental factors, such as background noise, physical obstructions, or other interfering sources, often coexist with deployed systems. These interference sources can vary significantly depending on the location, usage context, and surrounding conditions. If such factors are not carefully accounted for during the system's design and testing phases, they can severely degrade the system's performance, leading to inaccuracies, reduced reliability, or even complete failure in achieving the desired functionality. Addressing these challenges requires robust design strategies, adaptive algorithms, and extensive testing across diverse scenarios to ensure consistent and reliable operation in real-world environments.
- **Limited programmability and configurability:** Commodity devices often offer limited programming interfaces for configuring operational parameters, manipulating waveforms, or recording precise timing, restricting their flexibility for advanced applications. Researchers and engineers must devise innovative workarounds to overcome these limitations, such as leveraging undocumented features, reverse engineering hardware behavior, or employing software-based techniques to approximate desired functionalities. These efforts are critical for adapting commodity devices to meet the demands of specialized tasks and for unlocking their potential in acoustic sensing applications.

1.2 Acoustic Front Ends on Commodity Devices

An acoustic sensing system consists of four key components: acoustic signal emitter(s) (i.e., speakers), acoustic signal receiver(s) (i.e., microphones), the transmission medium, and objects (animate or inanimate) that interact with the acoustic waveforms in the medium. These objects can be either the targets to be sensed or elements of the environment that impact signal propagation. Depending on the placement of the speakers and microphones, and drawing terminologies from radar systems [23], an acoustic sensing system can operate in *mono-static*, *bistatic*, or *multi-static* configurations (Fig. 1.1). In mono-static systems, also known as *acoustic sonars*, speakers and microphones are colocated on the same device and typically driven by the same clock. These systems are simple and easy to deploy

Fig. 1.1 Different configurations of acoustic sensing systems. Left: mono-static; Middle: bistatic; Right: multi-static

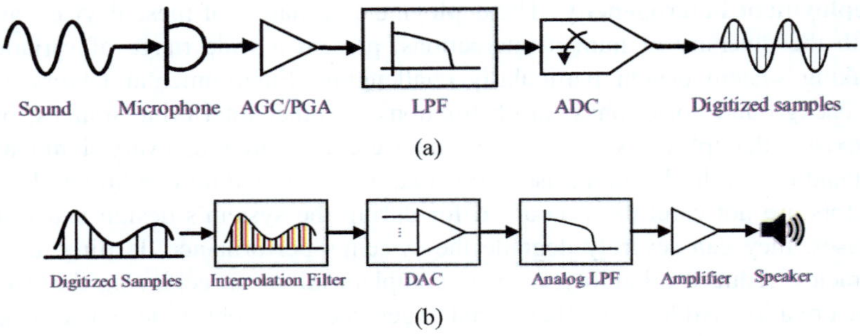

Fig. 1.2 Diagrams for typical acoustic hardware. (**a**) Sound recording system. (**b**) Sound playback system

but have a limited operational range, usually confined to the close proximity of the device. Bistatic acoustic systems have separate devices for transmitting and receiving acoustic signals, while multi-static systems involve multiple transmitters and/or receivers spatially separated at different locations. Deploying bistatic and multi-static systems is more complex. These systems can operate asynchronously [2, 31] or need to be synchronized through other modalities [21]. Their main advantages are increased coverage and enhanced spatial resolution when multiple units are deployed.

A typical pipeline of *sound recording and playback systems* is shown in Fig. 1.2a and b, respectively. The former converts mechanical (acoustic) waves into digital samples, while the latter does the opposite. In a recording system, acoustic signals are first converted into voltage signals by a *microphone*. An automatic gain control (AGC) or programmable gain amplifier (PGA) then amplifies the voltage signals to fit the dynamic range of a posterior analog-to-digital converter (ADC); this helps to improve the digitization resolution and to avoid saturation [19]. Amplified signals further go through a low-pass filter (LPF), also known as an anti-aliasing filter, and become band-limited signals. The cutoff frequency of the LPF is typically set to $f_s/2$, where f_s is the sampling rate. Filtered signals are finally converted into digital samples by an ADC. A sound playback system reverses the above process. Digital samples are first interpolated and then fed into a digital-to-analog converter (DAC) to become analog signals. The analog signals are further amplified and finally converted into acoustic waves by a *speaker*.

On acoustic playback systems, amplification or gains of sounds can often be configured. High-power speakers are utilized in applications that operate at long ranges. The most important parameters for acoustic recording systems include sampling rate, bit resolution, and the number of channels (or microphones). Higher sampling rates and bit resolutions can improve signal quality, allowing for greater frequency content and reduced quantization noise. Additionally, increasing the number of microphone channels enhances spatial resolution, enabling more detailed acoustic representations. In commodity systems, sampling rates typically

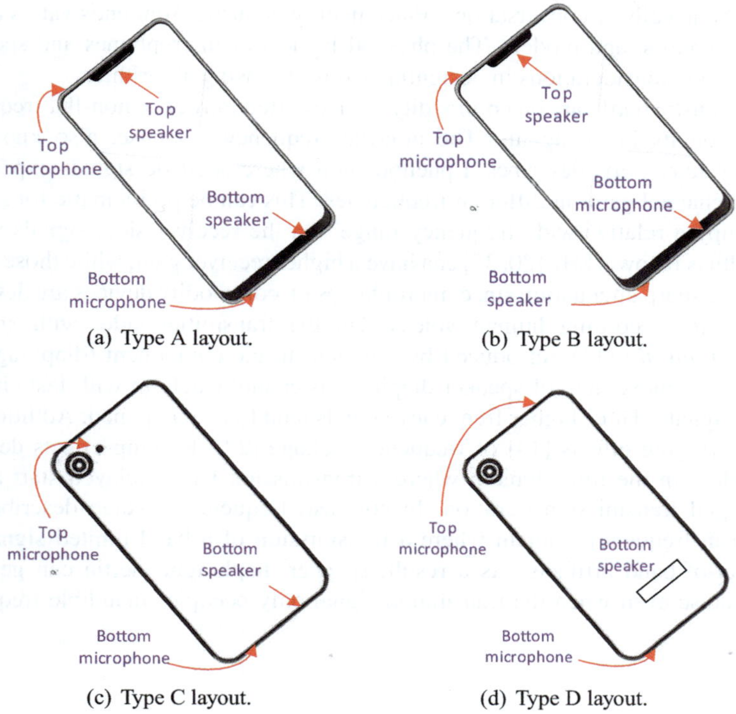

Fig. 1.3 Acoustic sensor layouts of different phone models: Front views on (**a**) and (**b**) of Type A (e.g., HUAWEI MATE 30 and Honor V20) and Type B (e.g., OPPO A59 and Xiao MI8 SE). Back views on (**c**) and (**d**) of Type C (e.g., One Plus 5T) and Type D (e.g., Samsung Galaxy S5 and HUAWEI MATE9)

range from 8 to 44.1 kHz, but certain devices (e.g., high-end smartphones) have a maximum sampling rate of 192 kHz [7]. The most common configuration for bit resolution is 8 bits or 16 bits, but customized devices may support up to 32 bits.

Contemporary commodity devices are often equipped with more than one microphone and one speaker. For instance, modern smartphones utilize two speakers to play stereo audio and employ two microphones to enhance recording qualities. The typical layouts of microphones and speakers on smartphones are shown in Fig. 1.3. Similarly, smart earbuds devices have multiple microphones to enhance audio quality and enable additional features such as noise cancelation. For example, in an Apple AirPods pro earbud, an external microphone detects external sound for noise cancelation and transparency mode. It helps filter out unwanted external noise while allowing important sounds, like voices, to be heard clearly when in transparency mode. An internal microphone facing the ear canal works in conjunction with the external microphone to optimize audio quality and noise cancelation; and an microphone at the bottom of the stem is primarily used for capturing user voice during phone calls or conversations. Placement of acoustic front ends varies across devices, vendors, and models. The physical layouts of microphones and speakers are important considerations in designing acoustic sensing algorithms.

The acoustic hardware on commodity devices often introduce non-flat frequency response on its input signals. The non-flat frequency response, also known as *frequency selectivity*, describes a phenomenon where acoustic signals experience different channel gains at different frequencies. This can be problematic for signals that occupy a relative wide frequency range. On the receiver side, signals whose bandwidth is below 8 kHz [20, 37] can have a higher receiver gain, while those above experience sharp attenuation since microphones on commodity devices are designed primarily for recording human voices. On the transmitter side, with *speaker diaphragm inertia* [17], introduced by a nonelectronic component (diaphragm) in speakers, the movement of speaker diaphragms cannot catch up with fast changes in input signals. Thus, higher frequency signals tend to be dampened. Additionally, it causes ringing effects [13] or frequency leakage [27]. Ringing effects describe the problem in the time domain where a transmission has a delayed start and/or a prolonged transmission duration. In contrast, frequency leakage describes the problem in frequency domain where a transmission of a band-limited signal can cause out-of-band artifacts. As a result, speaker diaphragm inertia can generate audible noise even when the transmitted signal only occupies inaudible frequency bands.[1]

1.3 Taxonomy of Acoustic Sensing Applications

The research community and industry have developed a diverse range of acoustic sensing applications, with new innovations emerging at an accelerating pace. Given the breadth and rapid evolution of this field, compiling an exhaustive list of all applications is neither feasible nor practical. Instead, this section provides a taxonomy of representative applications to illustrate the versatility of acoustic sensing and to highlight the shared underlying principles that enable these systems.

Existing applications can broadly be categorized into three main groups as summarized in Fig. 1.4. The first category, *aerial acoustic communication* (AAC), involves the use of sound waves to transmit data through the air, enabling low-power communication systems. Examples include device pairing, short-range data transfer, and authentication mechanisms using sound.

The second category, *location services*, leverages acoustic signals for spatial awareness. These applications include indoor localization, distance estimation, and object tracking. By analyzing the propagation, reflection, and interference

[1] Most people normally do not hear sounds about 18 kHz [3, 11, 18]. Therefore, the frequency range beyond 18 kHz is often deemed *inaudible* and hence commonly used for acoustic communication and sensing.

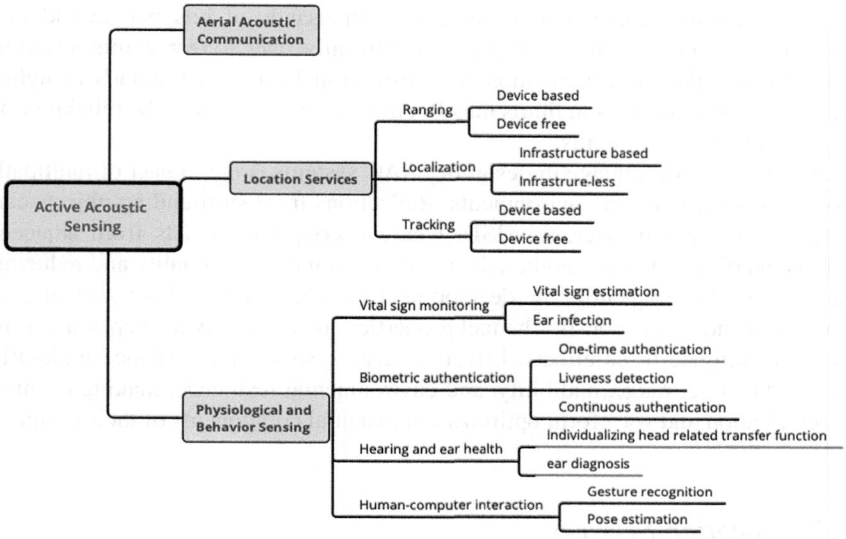

Fig. 1.4 Taxonomy of acoustic sensing applications

patterns of sound waves, such systems can provide accurate location information in environments where traditional methods, such as global positioning systems (GPS), are unavailable or unreliable.

The third category, *physiological and behavioral sensing*, uses acoustic signals to monitor human activities, health, and behaviors. Applications in this group range from estimating heart and respiratory rates to identifying gestures, postures, and diagnosing hearing disorders. These systems rely on the interaction between sound waves and the human body to extract meaningful information noninvasively.

1.3.1 Aerial Acoustic Communication

AAC enables devices equipped with embedded microphones and speakers to communicate without the need for additional hardware or complex network configurations. This approach offers a practical alternative to traditional RF-based device-to-device communication technologies, such as Bluetooth, near-field communication (NFC), and WiFi Direct. AAC achieves communication by encoding information into variations of acoustic waveforms. As discussed in detail in Chap. 4, different AAC implementations primarily differ in the waveforms and bandwidths they utilize, which in turn influence their achievable data rates, operational ranges, and levels of audibility.

The data rates of AAC systems are inherently constrained by the limited acoustic bandwidth available on commodity devices. Typically, such systems can achieve data rates ranging from hundreds to tens of thousands of bits per second [4]. However, their throughput can degrade significantly over longer communication distances due to the rapid attenuation of acoustic signals in air, particularly at higher frequencies. This attenuation limits the effective range and reduces the reliability of high-frequency AAC systems.

Another critical challenge in designing AAC systems is the impact of multipath propagation in real-world environments. Reflections from surrounding objects can create inter-symbol interference (ISI), where overlapping signals from adjacent symbols interfere with one another, degrading communication quality and reducing data integrity. Consequently, the development of robust AAC schemes requires a deep understanding of acoustic channel properties and the ability to adapt waveform designs to mitigate these effects. Effective AAC systems must balance trade-offs between data rate, range, audibility, and environmental resilience, making channel characterization and waveform optimization essential components of their design.

1.3.2 Location Services

Acoustic location services aim to determine the proximity of targets and localize and track them acoustically over time. With few exceptions, these applications typically rely on estimating the timing of acoustic waves (reflected or directly received) and leveraging the known propagation speed of sound in the medium to calculate distances or angles.

Existing work in this area can be further categorized into *device-based* and *device-free* approaches, as well as *infrastructure-based* and *infrastructure-less* solutions. In device-based approaches, the target is an IoT device programmed to emit acoustic signals, which can then be detected and localized using the received signals on other devices. These applications require the active cooperation of the target device. In contrast, device-free approaches involve passive targets and achieve localization and ranging by analyzing reflected acoustic signals, similar to sonar systems. In this case, acoustic signals emitted from a transmitter must travel twice the distance compared to device-based approaches, as they need to reach the target and return to the receiver. The effectiveness of device-free methods depends heavily on the material properties of the target, such as reflective coefficients, surface roughness, and dimensions, which influence the strength of the reflected signals. Consequently, device-free approaches often have more limited coverage and are more sensitive to environmental factors than their device-based counterparts.

The key difference between infrastructure-free and infrastructure-based approaches is whether custom-built devices are deployed within the environment to enable localization and tracking services. Infrastructure-free approaches often leverage existing environmental acoustic signatures to localize and track targets. While these solutions reduce hardware costs, they often require labor-intensive site surveys to generate location-dependent signal profiles, making them highly sensitive to environmental changes. In contrast, infrastructure-based approaches achieve higher localization accuracy by deploying dedicated hardware, such as acoustic anchors, throughout the environment. However, these systems come with added costs and complexities, including the need for precise synchronization and the logistical challenges of deploying multiple devices. As a result, although infrastructure-based solutions offer high accuracy, their hardware costs and pre-deployment efforts impose a significant barrier to widespread adoption.

1.3.3 Physiological and Behavioral Sensing

Physiological sensing involves measuring physiological parameters of the human body such as vital signs and anthropometric measurements, as well as for diagnosing health conditions. The prevalence of acoustic transducers in wearable devices, home, and vehicle environments offers new opportunities for physiological sensing without specialized devices. As an example, Fig. 1.5 reproduced from [5] illustrates how to detect mid-ear infection with fluid buildup using a smartphone. Behavior sensing, on the other hand, is broadly associated with monitoring, analyzing, and inferring human behaviors, activities, or interactions. Applications in this category are quite diverse, but they share similar operating principles, in estimating motion, deformation, and the presence or absence of objects based on the changes induced in acoustic channel characteristics.

Next, we discuss a few representative applications in physiological and behavioral sensing using acoustic signals.

Vital Sign Estimation

Vital signs, such as heart rate and respiration rate, provide critical insights into a person's overall health, physiological status, and potential medical conditions. Acoustic sensing offers a noncontact method for measuring vital signs, minimizing

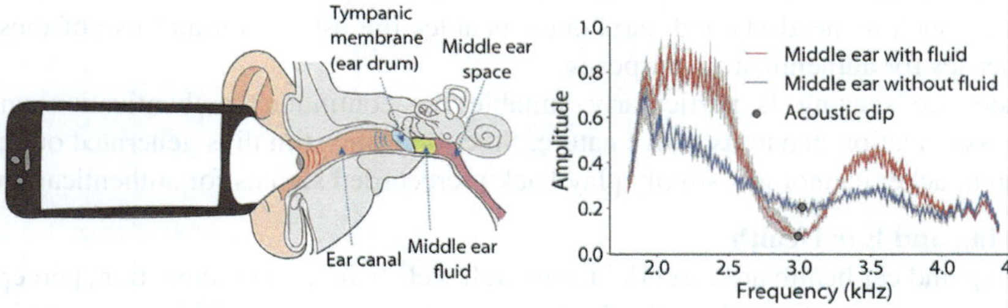

Fig. 1.5 Acoustic sensing can be used to detect mid-ear fluid. [Left] A cone-shaped paper funnel connects the bottom speaker of a smartphone on end and the ear canal on the other. [Right] Raw acoustic waveforms are captured by the smartphone microphone when chirps are played into ear with middle ear fluid (red) and without fluid (blue) (reproduced from [5])

discomfort and reducing the risk of skin irritation or infection, particularly for sensitive populations like newborns or burn victims. This technology enables continuous, real-time monitoring from a distance, making it well suited for applications in telemedicine and smart homes. Additionally, by estimating respiration rates, acoustic sensing can support human detection and help prevent unsafe situations, such as leaving small children unattended in vehicles.

Rhythmic chest movements caused by respiration and heartbeats introduce variations in the travel time of reflected acoustic waves from the human chest. The millimeter-level maximum displacements (approximately 12 mm from respiration and around 5 mm from heartbeats [6]) pose significant challenges for accurate timing estimation on commodity devices. Additionally, isolating motion induced by heartbeats from respiration presents a challenge, as respiratory movements tend to dominate. Furthermore, large body movements can introduce noise and potentially contaminate the estimates.

Biometrics Authentication

Biometrics authentication refers to the process of verifying a person's identity based on their unique biological or behavioral characteristics. It can be achieved using physical traits such as fingerprints, facial features, iris patterns, or voice, as well as behavioral traits like typing patterns or gait. Among authentication schemes, *one-time authentication* verifies a user's identity once, typically at the start of a session, whereas *continuous authentication* continually checks for signs that the authenticated user is still the one interacting with the system. A related concept is *liveness detection*, which determines whether the biometric sample being presented is from a live human being rather than from a fake or spoofed source.

Acoustic sensing can extract unique, subject-specific characteristics for authentication. For instance, the human ear canal is uniquely shaped, with diverse reflective and absorptive surfaces that generate distinctive echo patterns in response to external acoustic stimuli [26]. Additionally, when acoustic waves are bone-conducted through the head, the speed of sound transmission varies across different parts of the skull, and signal frequencies are attenuated differently, producing unique acoustic signals specific to each individual. The widespread availability of earable devices, such as headsets and earphones, enables the estimation and use of these biometrics for authentication purposes.

Acoustic sensing is particularly suitable for continuous authentication and liveness detection due to its active nature. Since acoustic stimuli is generated on the fly, an attacker cannot just simply play back prerecorded signals for authentication.

Hearing and Ear Health

Hearing and ear health are essential for overall well-being, communication, perception, safety, and quality of life. Head-related transfer functions (HRTFs) characterize the directional and frequency-dependent filtering effects of human body, especially the head, pinna, and torso [9]. Sounds coming from a point in space are subject to this filtering effect before reaching one's ear canals, making the perception of

3D sound locations possible. Since HRTFs are closely related to the size and the shape of head and ears, they are subject dependent and are defined separately for each ear at different azimuth and elevation directions. Obtaining an individual's HRTFs is important in a variety of hearing applications. For example, they allow accurate rendering of spatial audio, ensuring sounds appear to originate from precise locations in a 3D environment, which find applications in virtual reality and augmented reality [36]. In hearing aids or assistive hearing devices, custom HRTFs optimize sound delivery, improving sound localization and speech perception for users with hearing impairment [28]. Traditionally, estimating HRTFs requires specialized equipment, such as anechoic chambers and speaker arrays. Developing alternative methods that allow users to personalize HRTFs using commodity devices is an active area of research.

Acoustic sensing can play a pivotal role in diagnosing hearing impairment and other ear-related conditions. For example, a mobile app that performs basic tests, such as pure tone audiometry and speech intelligibility tests (e.g., digits in noise), enables early detection of hearing impairment from the convenience of home. Additionally, as illustrated in Fig. 1.5, a simple setup using a smartphone and a paper funnel is effective in detecting middle ear fluid, making it an appealing and accessible screening tool for parents and healthcare providers, particularly in resource-limited regions such as rural areas or the global south.

Human-Computer Interactions

Gesture interaction plays a significant role in human-computer interaction (HCI), offering an intuitive and natural way for users to communicate with technology. This form of interaction leverages the movement of hands, fingers, and sometimes the entire body to convey commands, manipulate objects, and navigate interfaces. Acoustic-based gesture interaction uses sound waves to detect and interpret gestures. There are generally two lines of approaches: gesture recognition and (hand) pose estimation.

Gesture recognition aims to understand the expressive meaning of body parts, in particular hands, serving as an interface for humans to interact with smart devices. The primary principle of these gesture recognition systems is to map appropriate acoustic channel state information to a finite set of performed gestures [8, 11, 14, 24, 32]. Exploiting trajectory information over time (e.g., Doppler profile) and language models, acoustic gesture recognition enables applications such as sign language understanding [10] and text entry [38].

Pose estimation aims to detect and track the positions of various key points on the human body, such as joints and anthropometric landmarks, to understand the configuration of the body or hand. In acoustic pose estimation, the inputs correspond to reflected acoustic signals from human body, and the outputs are key joint positions and/or joint angles [25, 33]. Compared to gesture recognition, pose estimation is not limited to a fixed set of gestures, making it instrumental in augmented reality, animation, and healthcare applications.

1.4 Organization

The preceding discussion highlights the diverse applications of acoustic sensing. Each category of applications requires tailored waveform design and processing techniques to meet specific requirements. However, a closer examination of existing solutions reveals shared building blocks and key techniques such as timing and channel estimations that are universally relevant.

The organization of the monograph is illustrated in Fig. 1.6. After the introduction, it proceeds with an exploration of the fundamental properties of aerial acoustic and structure-borne channels (Sect. 2.1) and introduces waveform designs tailored to each channel along with key design considerations (Chap. 2). It then delves into the core building blocks essential for extracting critical measurements, such as timing, direction, and channel responses (Chap. 3). These foundational techniques enable a wide range of applications, including AAC (Chap. 4), location services (Chap. 5), and physiological and behavioral sensing (Chap. 6). In Chaps. 4–6, we will present state-of-the-art methods and results in the respective applications.

Acoustic sensing is a rapidly advancing field, especially with the rise of embodied artificial intelligence (AI). As AI systems become more integrated with the physical world, acoustics continues to be a critical modality. Although this monograph cannot cover every potential application of acoustic sensing, the core techniques and building blocks presented for representative use cases will remain relevant and provide a solid foundation for future developments in the field.

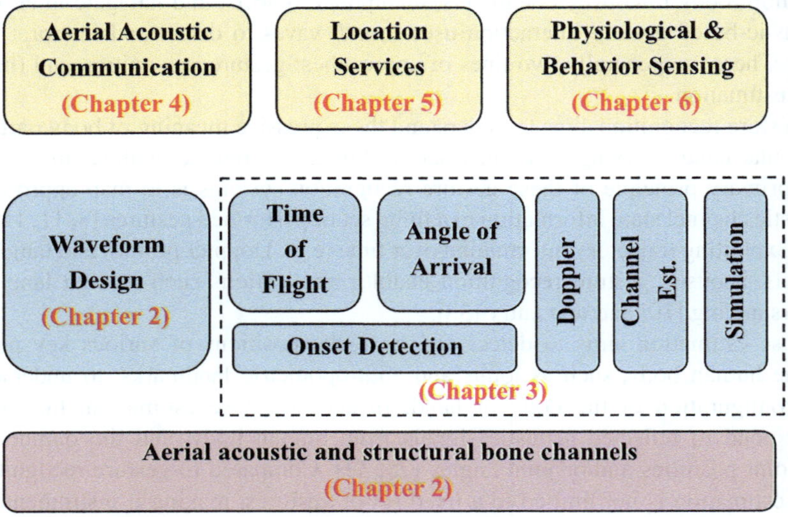

Fig. 1.6 Overall organization

References

1. F. Adib et al., 3D tracking via body radio reflections, in *Proceedings of the 11st USENIX NSDI* (2014), pp. 317–329
2. C. Cai et al., Asynchronous acoustic localization and tracking for mobile targets. IEEE Int. Things J. **7**, 830–845 (2019)
3. C. Cai et al., AcuTe: Acoustic thermometer empowered by a single smartphone, in *Proceedings of the 18th ACM SenSys* (2020), pp. 28–41
4. C. Cai, R. Zheng, J. Luo, Ubiquitous acoustic sensing on commodity iot devices: A survey. IEEE Commun. Surv. Tutor. **24**(1), 432–454 (2022)
5. J. Chan et al., Detecting middle ear fluid using smartphones. Sci. Translat. Med. **11**(492), eaav1102 (2019)
6. A. De Groote et al., Chest wall motion during tidal breathing. J. Appl. Physiol. **83**(5), 1531–1537 (1997)
7. *Google Pixel.* https://store.google.com/ (2021)
8. S. Gupta et al., SoundWave: Using the doppler effect to sense gestures, in *Proceedings of the 30th ACM SIGCHI* (2012), pp. 1911–1914
9. K. Iida. *Head-related transfer function and acoustic virtual reality* (Springer, Singapore, 2019). ISBN: 978-98-11397-45-5
10. Y. Jin et al., Sonicasl: An acoustic-based sign language gesture recognizer using earphones, in *Proceedings of the ACM on Interactive, Mobile, Wearable and Ubiquitous Technologies* , vol. 5, no. 2 (2021), pp. 1–30
11. S. Ke et al., VSkin: Sensing touch gestures on surfaces of mobile devices using acoustic signals, in *Proceedings of the 24th ACM MobiCom* (2018), pp. 591–605
12. S. Kumar et al., Accurate indoor localization with zero start-up cost, in *Proceedings of the 20th Annual International Conference on Mobile Computing and Networking (MobiCom)*. MobiCom '2014 (2014)
13. H. Lee et al., Chirp signal-based aerial acoustic communication for smart devices, in *Proceedings of the 34th IEEE INFOCOM* (2015), pp. 2407–2415
14. K. Ling et al., UltraGesture: Fine-grained gesture sensing and recognition, in *Proceedings of the 15th IEEE SECON* (2018), pp. 1–9
15. C. Liu et al., Lasagna: Towards deep hierarchical understanding and searching over mobile sensing data, in *Proceedins of the 22nd ACM MobiCom* (2016), pp. 334–347
16. F. Liu et al., Integrated sensing and communications: Toward dual- functional wireless networks for 6G and beyond. IEEE J. Sel. Areas Commun. **40**(6), 1728–1767 (2022)
17. Loud Speaker Diaphragm. *US Patent 2,905,260* (1959)
18. W. Mao, J. He, L. Qiu, CAT: High-precision acoustic motion tracking, in *Proceedings of the 22nd Annual International Conference on Mobile Computing and Networking* (2016), pp. 69–81
19. *MAX9814 Datasheet.* https://datasheets.maximintegrated.com/en/ds/MAX9814.pdf (2019)
20. R. Nandakumar et al., Dhwani: Secure peer-to-peer acoustic NFC, in *Proceedings of the ACM SIGCOMM 2013 Conference on SIGCOMM*. SIGCOMM '13 (2013)
21. N.B. Priyantha, The cricket indoor location system. PhD thesis. Massachusetts Institute of Technology, 2005
22. Q. Pu et al., Whole-home gesture recognition using wireless signals, in *Proceedings of the 19th ACM MobiCom* (2013), pp. 27–38
23. M.A. Richards et al., *Fundamentals of radar signal processing*. vol. 1 (McGraw-Hill, New York, 2005)
24. W. Ruan et al., AudioGest: Enabling fine-grained hand gesture detection by decoding echo signal, in *Proceedings of the 18th ACM UbiComp* (2016), pp. 474–485
25. Y. Shibata et al., Listening human behavior: 3d human pose estimation with acoustic signals, in *Proceedings of the IEEE/CVF Conference on Computer Vision and Pattern Recognition* (2023), pp. 13323–13332

26. M.R. Stinson, B.W. Lawton, Specification of the geometry of the human ear canal for the prediction of sound-pressure level distribution. J. Acoust. Soc. Am. **85**(6), 2492–2503 (1989)
27. Y.-C. Tung, K.G. Shin, Expansion of human-phone interface by sensing structure-borne sound propagation, in *Proceedings of the 14th ACM MobiSys* (2016), pp. 277–289
28. J. Usher, W.L. Martens, Perceived naturalness of speech sounds presented using personalized versus non-personalized HRTFs, in *13th In- ternational Conference on Auditory Display, Montreal, Canada* (2007)
29. D. Vasisht, S. Kumar, D. Katabi., Decimeter-level Localization with a single WiFi access point, in *Proceedings of the 13th USENIX NSDI* (2016), pp. 165–178
30. W. Wang et al., Understanding and modeling of WiFi signal based human activity recognition, in *Proceedings of the 21st ACM MobiCom* (2015), pp. 65–76
31. Y.-T. Wang et al., ARABIS: an asynchronous acoustic indoor positioning system for mobile devices, in *2017 IEEE International Conference on Indoor Positioning and Indoor Navigation (IPIN 2017)* (2017)
32. Y. Wang, J. Shen, Y. Zheng, Push the limit of acoustic gesture recognition, in *Proceedings of IEEE INFOCOM* (2020), pp. 566–575
33. S. Wang et al., Towards smartphone-based 3D hand pose reconstruction using acoustic signals. ACM Trans. Sen. Netw. **20**, 1–32 (2024)
34. Y. Yang et al., A survey on security and privacy issues in internet-of-things. IEEE Int. Things J. **4**(5), 1250–1258 (2017)
35. Y. Yang, J. Nie, J. Luo., ReflexCode: Coding with super-posed reflection light for LED-camera communication, in *Proceedings of the 23rd Annual International Conference on Mobile Computing and Networking*. MobiCom '17 (2017)
36. N.H. Zandi, A.M. El-mohandes, R. Zheng, Method and system for determining individualized head related transfer functions. US Patent App. 18/580,344 (2024)
37. B. Zhou et al., BatMapper: Acoustic sensing based indoor floor plan construction using smartphones, in *Proceedings of the 15th ACM MobiSys* (2017), pp. 42–55
38. Y. Zou et al., EchoWrite: An acoustic-based finger input system without training, in *Proceedings of IEEE INFOCOM* (2019), pp. 778–787

Chapter 2
Acoustic Waveform Design

2.1 Aerial- and Structure-Borne Channels

Acoustic signals, emanating from mechanical vibrations, possess the unique ability to traverse various media, including gases, liquids, and solids. In this monograph, we concentrate on acoustic sensing facilitated by commodity devices. Consequently, the discussion focuses on aerial- and structure-borne signals, along with the respective channel attributes.

2.1.1 Physics and Propagation Characteristics

Acoustic signals, fundamentally mechanical vibrations, are generated by a wide range of excitation sources. These include the oscillating diaphragm of a speaker, human vocal cords, the vibrating strings and sound board of instruments like guitars and violins, an operating engine, or even a bottle striking the floor. As long as the excitation source continues to vibrate, the resulting acoustic signals propagate through the surrounding medium.

Acoustic signals propagating through air, also termed airborne sound, exhibit a relatively constant velocity of approximately 340 meters per second under standard atmospheric conditions. These signals are generally considered nondispersive, meaning they tend to maintain their original waveform during propagation. Conversely, acoustic signals traveling through solid media, known as structure-borne signals, typically experience dispersion. In such media, the phase velocity is frequency dependent, with higher frequency components often propagating at a greater velocity than lower frequency components, leading to temporal spreading of the waveform, as illustrated in Fig. 2.1.

Despite the possibility of a shared origin in an excitation source, airborne and structure-borne signals exhibit fundamentally different propagation behaviors due to

R. Zheng, C. Cai, *Acoustic Sensing on Commodity Devices and its Applications*, Wireless Networks, https://doi.org/10.1007/978-3-031-96875-4_2

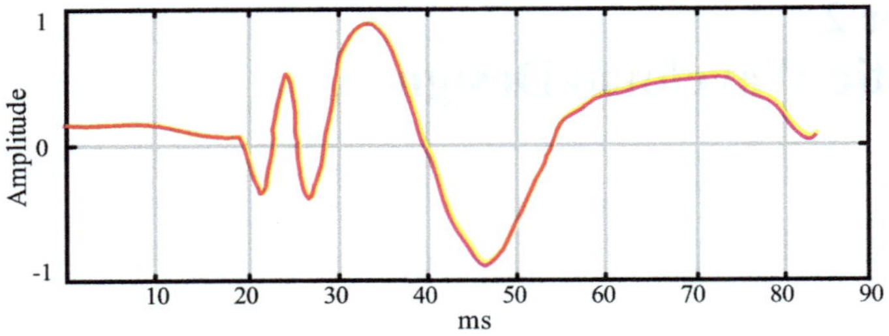

Fig. 2.1 Acoustic dispersion during propagation in solid media [36]. The waveform is recorded by an earbud (with on-board microphone) when a user taps a tooth

the underlying physical principles, thereby yielding distinct channel characteristics. Given this crucial differentiation, the subsequent discussion will focus on the channel properties associated with both air and solid media.

2.1.1.1 Airborne Channel

Acoustic signals propagating through air typically emanate from the source as expanding spherical wavefronts. This propagation mechanism is a result of the cyclical compression and rarefaction of air molecules, which serve to transmit pressure variations through the medium. As these wavefronts expand radially, the acoustic signals undergo gradual attenuation due to geometric spreading, atmospheric absorption, and other dissipative processes, ultimately leading to a reduction in energy and a consequent decrease in signal amplitude.

In airborne acoustic channels, signal propagation occurs without a physically discernible medium, a characteristic that often invites comparison to wireless electromagnetic signals, such as radio waves [10]. These signals are subject to several phenomena, including path loss, where signal intensity diminishes with increasing distance from the source, and the Doppler effect, which describes the apparent change in signal frequency due to the relative motion between the source and the receiver. Furthermore, acoustic waves in air can experience reflection upon encountering surfaces, refraction when traversing regions of differing air density, diffraction as they bend around obstructions, and scattering upon interaction with complex structures. The confluence of these effects significantly impacts both the propagation range and the fidelity of airborne acoustic signals.

Path Loss

Path loss constitutes a fundamental concept in wireless communication, quantifying the reduction in signal power density as it propagates through space [43]. This same principle governs the behavior of acoustic signals in air, where energy is dissipated as the distance from the source increases. Under ideal free-field conditions, acoustic intensity adheres to the inverse-square law, indicating that it decreases proportionally to the square of the distance from the source:

$$p_d \propto \frac{p_0}{r^2}, r \geq r_0, \tag{2.1}$$

where p_0 denotes the acoustic intensity at a reference distance r_0 (e.g., $r_0 = 1$ m), and p_d is the intensity measured at a distance of r away from a sound source.

Equation (2.1) describes the geometric spreading of acoustic energy in a free-field scenario, predicated on the assumptions of isotropic radiation and the absence of reflections. However, in practical settings, especially within enclosed environments, acoustic signals commonly experience multipath propagation, wherein reflections from surfaces can lead to constructive or destructive interference. Consequently, the idealized free-field assumption often proves inadequate, and empirically observed signal levels frequently exhibit substantial deviations from the model's predictions.

Doppler Effect

The Doppler effect, also known as the Doppler shift, is a phenomenon described in [4] wherein the perceived frequency of a wave undergoes a change for an observer in motion relative to the wave's source. First elucidated by the Austrian physicist Christian Doppler in 1842, this fundamental principle finds application across a wide range of scientific and technological domains, including acoustics, astronomy, and radar technology.

When a source and an observer are in relative motion toward each other, successive wave crests are emitted from positions of decreasing separation relative to the observer, resulting in an apparent increase in the wave frequency. Conversely, as the source and observer recede from one another, wave crests originate from progressively more distant locations, leading to a perceived decrease in frequency. A familiar example of this effect is the alteration in the perceived pitch of a vehicle's horn as it approaches and subsequently moves away from a stationary observer. As illustrated in Fig. 2.2, the observed frequency is maximal during the approach, equals the source frequency at the instant of passing (due to zero radial velocity), and subsequently diminishes as the separation increases.

For acoustic signals propagating through air, the velocities of both the observer and the source are defined relative to the medium (air) in which the waves travel. As a result, the total Doppler effect can arise from the motion of the source, the observer, the medium, or any combination of these factors. Specifically, as described in [16], the frequency shift is given by

$$\Delta f = \frac{\Delta v_d}{c} f, \tag{2.2}$$

where f is the original frequency, Δv_d denotes the relative radial velocity between the sound source (transmitter) and the observer (receiver), and c is the speed of sound in air.

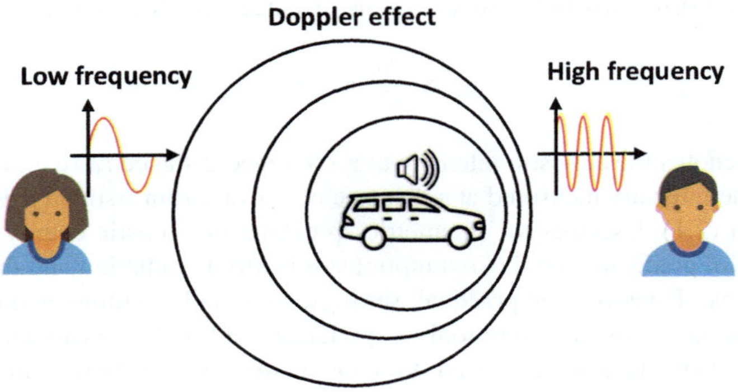

Fig. 2.2 Acoustic Doppler effect

This equation provides a first-order approximation of the Doppler shift, assuming low relative velocities compared to the speed of sound and no medium-induced effects (e.g., wind or temperature gradients).

Temperature Effect
The influence of thermodynamic conditions on wave propagation speed, a phenomenon often termed the temperature effect [32], is a fundamental physical principle observed in both gaseous and solid media. While a comprehensive analysis of this effect can be intricate, our focus herein is on its specific impact on acoustic signals in air. Within this context, the speed of sound is primarily governed by temperature and moisture content, as mathematically expressed by

$$c = \sqrt{403T_p\,(1 + 0.32e/\rho)}, \tag{2.3}$$

where c represents the sound speed in meters per second (m/s) in air, T_p is the absolute temperature in Kelvin (K), e denotes the vapor pressure of water in air, and ρ signifies the absolute atmospheric pressure (typically measured in Pascals, atmospheres, or millibars). As evident from Eq. (2.3), the speed of sound exhibits a dependence on humidity through the term e/ρ. However, given that this ratio is generally small, a common approximation for the speed of sound in air, particularly when humidity effects are secondary, can be expressed as $c^2 = 403T_p$.

Reflection, Diffraction, and Scattering
Interactions of acoustic waves with boundaries are governed by fundamental principles of wave physics, notably reflection, diffraction, and scattering. Reflection occurs when an acoustic wave encounters an interface between two media, resulting in a portion of the wave propagating back into the originating medium. In the context of airborne sound, this phenomenon is commonly observed as sound waves reflecting from rigid surfaces, such as walls or ceilings, as depicted in Fig. 2.3. A familiar manifestation of reflection is the echo perceived in large, unoccupied concert halls, where sound waves produced by a vocalization reflect off distant

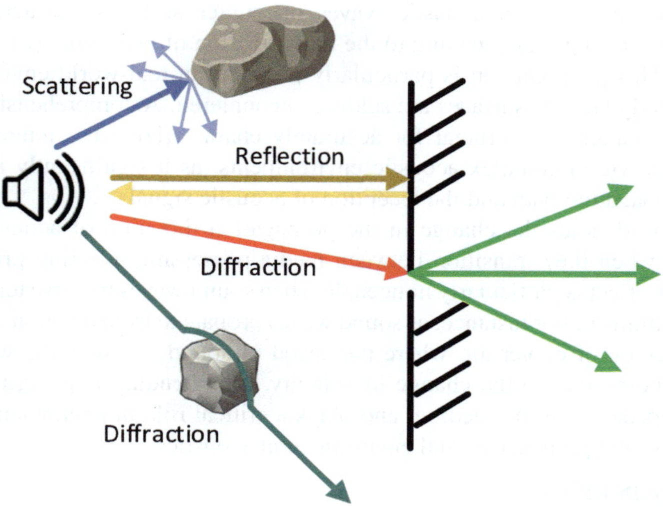

Fig. 2.3 Acoustic propagation model in air channel

boundaries and return to the listener after a brief temporal delay, creating a discernible repetition of the original sound.

The intensity and directionality of reflected acoustic signals are determined by the acoustic impedance mismatch between the interfacing media and generally adhere to the laws of reflection analogous to those in optics. In practical scenarios, acoustic energy often disperses across multiple angles, particularly when emanating from an omnidirectional source, such as a typical loudspeaker. Consequently, when an acoustic wave impinges upon the surfaces of an object, reflections of varying intensities can be detected across a spectrum of directions. Recent investigations [41, 46] suggest that acoustic reflections can exhibit sparsity, with significant signal intensity concentrated in a limited number of directions. To enhance directional control and focus acoustic propagation, techniques such as beamforming utilizing acoustic arrays and the application of acoustic metasurfaces [49] have been explored. These methodologies can effectively improve signal intensity and the signal-to-noise ratio (SNR) in a targeted direction, rendering them valuable for applications demanding precise sound localization and energy concentration.

Diffraction describes the phenomenon of wave bending and spreading that occurs when waves encounter an aperture or an obstacle. This effect becomes most prominent when the characteristic dimensions of the obstacle or the opening are comparable to or smaller than the wavelength of the incident wave. As illustrated in Fig. 2.3, even when acoustic signals (represented by the red ray) are emitted as a narrow beam, their passage through a constricted aperture results in a divergence of the beam on the distal side. Furthermore, acoustic waves exhibit the capacity to bend around obstacles, enabling sound propagation into regions that would otherwise lie within a geometric shadow zone.

Scattering arises when acoustic waves encounter surfaces characterized by irregularities or roughness, leading to the redistribution of wave energy in multiple directions. This phenomenon is particularly pertinent in real-world environments, where perfectly smooth surfaces are seldom encountered. A comprehensive understanding of scattering is crucial for accurately characterizing the interactions of sound waves within complex acoustic environments, as it significantly influences both the propagation path and the reception of acoustic signals.

Refraction denotes the change in the propagation direction of acoustic waves that occurs when they transition between media possessing differing propagation speeds. This effect is particularly noticeable when sound waves traverse temperature gradients within air. For instance, as sound waves propagate from a region of warmer air into a region of cooler air, where the speed of sound is lower, the wavefronts undergo bending due to the change in velocity. This bending is governed by the acoustic impedance of the medium and plays a critical role in phenomena such as long-range sound propagation and environmental acoustics.

The Multipath Effect

The multipath effect in acoustic signal propagation is a well-established phenomenon wherein sound waves reach a receiver via multiple distinct paths. This occurs due to reflections, refraction, diffraction, and scattering from surfaces and obstacles present in the propagation environment. These multiple propagation paths introduce variations in signal arrival times, phase shifts, and amplitude fluctuations at the receiver, resulting in either constructive or destructive interference.

As depicted in Fig. 2.4, when an acoustic device emits signals from a source location (A), these signals propagate radially in multiple directions, analogous to optical rays. If a receiver, such as a microphone, is positioned at point B, the received signals may arrive through various paths. Specifically, line-of-sight (LoS) signals refer to those propagating directly from A to B without encountering any intervening obstacles. These signals typically exhibit the highest energy due to minimal propagation loss and the shortest transit time. Conversely, non-line-of-sight (NLoS) signals, also known as multipath signals, encompass those that undergo one or more reflections from surfaces such as points C or D before reaching B. Owing to the additional propagation losses associated with reflections, these signals generally possess lower energy compared to LoS signals.

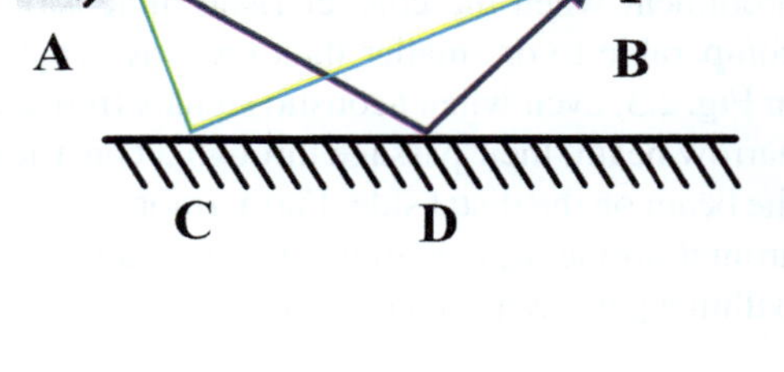

Fig. 2.4 The multipath effect

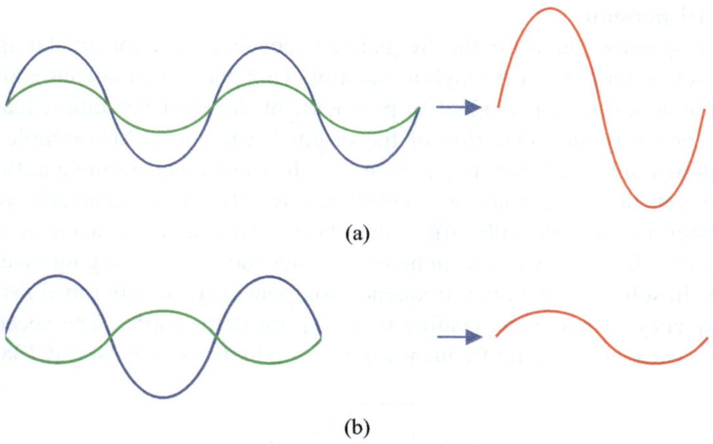

Fig. 2.5 Multipath signals can reinforce or cancel each other, depending on the phase difference between each component. (a) In-phase signals add up constructively. (b) Out-of-phase signals add up destructively

However, determining the relative energy levels of LoS and NLoS signals presents a nontrivial challenge. While LoS signals are often assumed to maintain a relatively stable energy level, NLoS signals can exhibit significant variations in energy depending on the signal frequency and the specific characteristics of the environment. When NLoS components arrive at the receiver in phase (i.e., their wave peaks are aligned), as illustrated in Fig. 2.5a, they undergo constructive interference, potentially resulting in a combined signal energy that exceeds that of the LoS component. Conversely, when NLoS components arrive out of phase, they interfere destructively, leading to a significant attenuation of the received signal compared to the energy of individual NLoS components.

The intricate interplay between multipath propagation, interference phenomena, and the resultant variations in signal strength underscores the characterization and mitigation of multipath effects as a critical area of focus in acoustic signal processing.

2.1.1.2 Structure-Borne Channel

Acoustic signals, being fundamentally mechanical vibrations, can propagate through solid media exhibiting unique characteristics that differ from airborne propagation. This section will examine two key phenomena associated with structure-borne channels: acoustic dispersion and acoustic resonance.

Acoustic Dispersion

Acoustic dispersion refers to the frequency-dependent variation in the speed of sound waves as they travel through a medium. This phenomenon causes different frequency components of a signal to propagate at different velocities, leading to temporal spreading and distortion of the original waveform. An example of this effect is shown in Fig. 2.1, where a person taps their teeth, generating a pulse vibration that travels through the jaw and skull bones, resulting in a dispersed waveform.

Dispersion can significantly affect the clarity and fidelity of acoustic signals, especially over long distances or in heterogeneous media with varying mechanical properties. In solid media, higher frequency components generally travel faster than lower frequency components, leading to waveform deformation. The speed c_f of an acoustic wave at a specific frequency f is given by Ross and Ostiguy [38]:

$$c_f = \sqrt[4]{\frac{Ehf^2}{12\rho\left(1 - v_p^2\right)}}, \tag{2.4}$$

where v_p is the phase velocity, and E, ρ, h are constants that characterize a medium. Specifically, E gives the elasticity, ρ describes the stiffness, and h represents the thickness of the medium.

Equation (2.4) highlights that both the material composition and the geometric structure of a solid medium significantly influence the propagation velocity of structure-borne signals. Therefore, medium properties play a crucial role in shaping the behavior of acoustic waves as they travel through solid materials.

Acoustic Resonance

Acoustic resonance [2] is a phenomenon wherein an acoustic system preferentially amplifies a specific frequency, diminishing others. This results in a strong response at that particular frequency, known as the *resonant frequency*. This selective amplification is crucial for manipulating sound and is a key principle in designing acoustic devices that control and direct sound waves. A thorough understanding and application of acoustic resonance are essential in acoustic engineering, where the aim is to optimize how sound waves interact with materials.

In a standard mass-spring configuration, as shown in Fig. 2.6a, we examine the behavior of a damped mass on a spring when an external sinusoidal force is applied. Using Newton's second law, the resulting motion can be described by the following equation [25]:

$$m\frac{d^2x}{dt^2} = F_0 \sin(\omega t) - kx - q\frac{dx}{dt}, \tag{2.5}$$

where m denotes the mass, x is the displacement of the mass from the equilibrium point, F_0 is the amplitude of the driving force, ω is the angular frequency, k is the spring constant, and q is the viscous damping coefficient. By defining the undamaged angular frequency $\omega_0 = \sqrt{k/m}$ and damping ratio $\eta = \frac{q}{2\sqrt{mk}}$, Eq. (2.5)

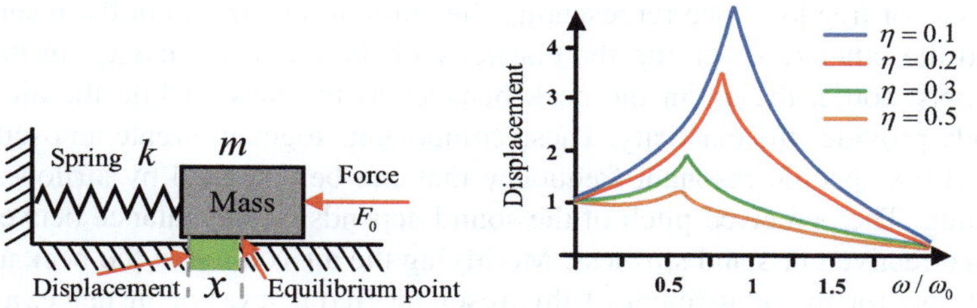

(a) A mass spring system.

(b) The amplitude or displacement over exciting frequency.

Fig. 2.6 (**a**) A mass-spring system model and (**b**) the displacement over exciting frequency

can be rewritten as follows:

$$\frac{d^2x}{dt^2} + 2\eta\omega_0\frac{dx}{dt} + \omega_0^2 x = \frac{F_0}{m}\sin(\omega t). \tag{2.6}$$

The comprehensive solution to the aforementioned equation encompasses both transient and steady-state components. The transient component diminishes rapidly, rendering the steady-state solution the primary focus for the analysis of resonance. The steady-state solution is characterized by the following expression:

$$x(t) = \frac{F_0}{m\sqrt{(2\omega\omega_0\eta)^2 + \left(\omega_0^2 - \omega^2\right)^2}}, \tag{2.7}$$

where $\phi = \arctan\left(\frac{2\omega\omega_0\eta}{\omega^2-\omega_0^2}\right) + n\pi(n = 0, 1, 2, ...)$. The amplitude of oscillation, which is the maximum displacement, exhibits considerable variation with changes in the damping ratio, as illustrated in Fig. 2.6b. Resonance is characterized by the attainment of the maximum displacement when the system's driving frequency coincides with its natural frequency. The resonant frequency ω_r can be obtained as $\omega_r = \omega_0\sqrt{1 - 2\eta^2}$.

It is important to note that acoustic resonance is not limited to solid materials. Various types of resonators, including those employing air as the medium, exist. Examples include Helmholtz resonators, quarter-wavelength resonators, and membrane-type resonators. A common illustration of a *Helmholtz resonator*, also known as a *cavity resonator*, is an empty soft drink bottle. The acoustic phenomenon of Helmholtz resonance occurs when air is blown across the opening of such a cavity. The resulting sound is characteristic of a lumped acoustic system, a simplified analytical approach where the system's properties are considered concentrated. This analysis often employs the long wavelength limit, which assumes the sound's wavelength is significantly larger than the resonator's dimensions. Under this assumption, the acoustic device can be modeled as a harmonic oscillator with two degrees of freedom: one representing the mass of the air within the resonator's neck and the other representing the elasticity of the air in the cavity. In the case of an empty bottle, the air in the neck behaves as the mass, while the air in the main body provides the elasticity. These components together create an oscillatory system with a specific resonant frequency that can be activated by airflow across the opening. The perceived pitch of the sound depends on the balance between the system's effective mass and stiffness. Modifying the air volume in the neck and the cavity allows for the adjustment of this resonant frequency. For instance, a wider opening leads to a higher frequency due to the decrease in the oscillating air mass. Conversely, increasing the volume or length of the neck results in a lower frequency because the larger air mass requires more energy to oscillate. The calculation of the resonant frequency for a cavity resonator can be generally expressed as

$$f_{\text{resonance}} = \frac{c}{2\pi}\sqrt{\frac{s}{vl}}. \tag{2.8}$$

In this equation, c represents the speed of sound, s is the area of the opening, v indicates the volume of the cavity, and l signifies its length. Comprehending these principles is instrumental in elucidating or developing specialized acoustic sensing technologies. Examples include the underlying physics of the occlusion effect within the auditory canal [18], or the design of specific resonators to selectively enhance acoustic signals modulated by biometric data for authentication purposes [48].

2.1.1.3 Nonlinearity of Acoustic Channels

Acoustic systems exhibit nonlinearity when the system's output deviates from a direct proportionality to the input. This behavior leads to various effects, including harmonic distortion, wave steepening, and intermodulation. This phenomenon can be observed in both airborne and structure-borne acoustic pathways, as well as in acoustic transducers like loudspeakers and microphones.

In airborne channels, nonlinearity becomes prominent with very intense sounds, such as those produced by jet engines or explosions, or with high-frequency signals like ultrasound. These conditions cause rapid fluctuations in air pressure, leading to distortions in the sound waves and the formation of shock waves. A notable application of this effect is in parametric loudspeakers, which utilize ultrasonic waves to generate a highly focused beam of audible sound through nonlinear interaction with the air. This makes them suitable for directional sound applications. In solid materials, nonlinearity can arise from excessively strong forces acting upon the material or from the presence of defects such as imperfections, cracks, or microfractures. Under these circumstances, sound waves do not propagate uniformly but experience distortion, the generation of harmonics, and the coupling of different vibrational modes. Nonlinearities in speakers and microphones stem from both mechanical and electrical imperfections, including nonlinearities in the

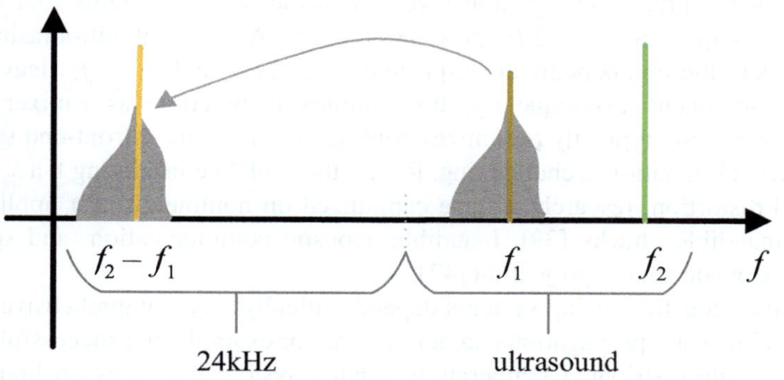

Fig. 2.7 The acoustic nonlinearity

diaphragm's movement, magnetic coil saturation, and distortions introduced by the amplifier circuitry.

Irrespective of its origins, nonlinearity within acoustic systems can result in the combination of signal components [39, 47]. Figure 2.7 provides an intuitive illustration of this phenomenon. Specifically, when two pure tone signals (in the ultrasound range) with frequencies f_1 and f_2 interact within an acoustic channel, they generate a novel frequency at the difference, $f_2 - f_1$. Mathematically, the resulting output y_{out} of an acoustic channel in response to an input x_{in} can be represented as:

$$y_{\text{out}} = \sum_{i=1}^{\infty} A_i x_{\text{in}}^i, \tag{2.9}$$

where A_i is the ith-order channel gain. When x_{in} contains two frequencies f_1 and f_2,

$$x_{\text{in}} = \sin(2\pi f_1 t) + \sin(2\pi f_2 t), \tag{2.10}$$

the output y_{out} becomes

$$y_{\text{out}} \left| \begin{aligned} &= \textstyle\sum_{i=1}^{\infty} A_i x_{\text{in}}^i \\ &= A_1 x_{\text{in}} + A_2(\sin(2\pi f_1 t) + \sin(2\pi f_2 t))^2 + \text{high order terms}. \end{aligned} \right. \tag{2.11}$$

Ignoring the gain, the second-order term in Eq. (2.11) can be expressed as

$$\begin{aligned} &[\sin(2\pi f_1 t) + \sin(2\pi f_2 t)]^2 \\ &= \sin^2(2\pi f_1 t) + \sin^2(2\pi f_2 t) + 2\sin(2\pi f_1 t)\sin(2\pi f_2 t) \\ &= 1 - \tfrac{1}{2}\cos(2\pi 2 f_1 t) - \tfrac{1}{2}\cos(2\pi 2 f_2 t) \\ &\quad + \cos(2\pi(f_1 - f_2)t) - \cos(2\pi(f_1 + f_2)t). \end{aligned} \tag{2.12}$$

Analysis of the preceding equation reveals that the output contains four distinct frequency components: $2f_1, 2f_2, f_1 + f_2, f_1 - f_2$. An internal anti-aliasing filter can eliminate the components at frequencies $2f_1, 2f_2$, and $f_1 + f_2$, leaving the $f_1 - f_2$ component. Consequently, this nonlinearity functions as a mixer. While careful engineering typically minimizes nonlinearity in acoustic front-end systems, its complete elimination is challenging. Rather than solely considering it a source of unwanted distortion, researchers have capitalized on nonlinearity for applications such as inaudible attacks [39], inaudible acoustic communication, and spatially directive acoustic source projection [47].

Effective acoustic sensing systems depend critically on a comprehensive understanding of acoustic propagation characteristics. For example, the successful design of a large-scale distributed temperature sensing system [7] relies on knowledge of how temperature affects the propagation of airborne acoustic signals. Similarly, identifying the influential factors of the dispersive phenomenon in structure-borne acoustic signals is crucial for developing an appropriate adversary training strategy for user-independent ranging [9]. In the following section, we will explore waveform design, examining waveforms specifically tailored for diverse applications and scenarios in acoustic sensing.

2.2 Waveforms for Acoustic Communication

While underwater acoustic communication has been extensively studied [27], this monograph centers on airborne acoustic communication, also known as aerial communication. This communication method facilitates a pervasive Internet of Things (IoT) ecosystem by enabling devices like smartphones, laptops, and smart speakers to establish secure, short-range channels for low-data-rate information exchange. The potential applications of this technology are vast. For instance, a smart TV could transmit supplementary information, such as a direct purchase link for a product featured in an advertisement, to a smartphone. Similarly, a smartphone could function as a seamless remote control for a smart TV, bypassing the complexities of traditional pairing and configuration methods like Bluetooth. The promise of these applications has spurred significant research into acoustic communication techniques. For example, major industry players, including Google, have adopted this approach to enable copresence among nearby devices [12].

The fundamental differences between radio frequency (RF) and acoustic wave propagation—particularly in signal velocity, attenuation characteristics, and bandwidth limitations—mean that the methodologies and techniques developed for RF communication cannot be directly applied to acoustic channels. Furthermore, many of the challenges encountered in RF communication are amplified in the realm of airborne acoustic communication. Specifically:

- *Doppler effect:* The slow propagation speed of sound (340 m/s at room temperature), coupled with the relatively low carrier frequencies used in airborne acoustic communication, means that even modest movements, such as human walking, can induce significant Doppler shifts. This poses a challenge for the design of resilient waveforms for mobile applications, requiring careful consideration of frequency shifts and temporal variations to guarantee reliable communication.

- *Vulnerability to ambient noise:* Airborne acoustic communication operates within an inherently restricted bandwidth, typically limited to the audible range of roughly 20 Hz–22 kHz. Conversely, ambient noise often possesses a wide spectral distribution, potentially covering the entire acoustic spectrum, which can severely impact signal quality. Thus, the design of waveforms for airborne acoustic communication necessitates the inclusion of effective noise reduction techniques to improve transmission reliability and preserve signal fidelity in noisy acoustic conditions.
- *Frequency selectivity:* Acoustic transducers, including speakers and microphones, are primarily designed for human speech, leading to an optimized frequency response around 4 kHz. Within this range, these transducers show increased gain and microphone sensitivity. However, outside this spectral region, both gain and sensitivity decrease considerably, resulting in significant frequency selectivity. This inherent nonuniform response of transducers requires the implementation of compensation techniques to minimize its effects on signal transmission and reception.

Despite these challenges, airborne acoustic communication offers several advantages over its radio frequency counterparts. Firstly, acoustic transducers are widely available in commodity devices and inherently support physical layer signal processing. This characteristic enables the implementation of various modulation schemes without requiring hardware modifications or facing compatibility issues. Secondly, the ratio of bandwidth to propagation speed for acoustic signals is significantly higher than that of radio frequency signals. This enhanced ratio improves the spatial resolution and sensitivity of acoustic-based sensing systems, allowing for more precise environmental characterization and object detection. Thirdly, airborne acoustic communication presents a lower risk of eavesdropping as acoustic signals have limited propagation range and do not easily penetrate walls.

To address the aforementioned challenges, waveforms for airborne acoustic channels are typically designed to be robust against Doppler shifts and ambient interference. The issue of frequency selectivity is commonly mitigated through the application of appropriate channel compensation techniques [11]. In the subsequent section, we introduce commonly used waveforms for aerial acoustic communication (AAC), including chirp signals [7] and their variants [8, 29], sinusoidal signals, and orthogonal frequency division multiplexing (OFDM) signals.

Chirp signals are among the most prevalent waveforms utilized in acoustic communication. They are commonly employed for encoding preambles during synchronization processes and can also be used to encode information symbols. Chirp signals exhibit strong resilience to background noise, significant Doppler effects, and multipath propagation, making them particularly attractive for challenging acoustic environments. Notably, their unique properties enable communication even below the noise floor. A chirp signal is characterized by a frequency that either increases (up-chirp) or decreases (down-chirp) over time. A specific type, the linear chirp, also known as a linear frequency modulated (LFM) signal or a swept-sine wave, maintains a constant rate of frequency change throughout its duration [34]. It is represented by

$$s(t) = a\cos\left(2\pi\left(f_{\min}t + \frac{k}{2}t^2\right) + \phi\right), 0 < t < T, \tag{2.13}$$

where $f_{\min}$ is the initial frequency, a is the maximum amplitude, ϕ is the initial phase, and $k = \frac{B}{T}$ is the *modulation coefficient* or *chirp rate* [29, 35, 40, 51]. The product of B and T denotes the *processing gain* of chirp signals [8]. The larger this value, namely $B \times T$, the better capability would chirp signals have in combating ambient interference. The time and time-frequency domain of chirp signals are shown in Fig. 2.8.

Various modifications of chirp signals have been employed in wireless communication to enhance performance under diverse channel conditions. Notable examples include exponential (or geometric) chirps [34], one- and two-dimensional chirps, chirps with random amplitude, and harmonic chirps [31]. However, some of these variations introduce significant computational complexity, rendering them impractical for resource-constrained IoT devices. Consequently, specialized chirp waveforms optimized for acoustic communication have been proposed.

To convey information, distinct information symbols need to be represented by different waveforms. Conventional approaches often utilize "orthogonal" symbols within a time division multiple access scheme. For instance, a dual-coefficient linear chirp was introduced in [29], as illustrated in Fig. 2.9a. By employing different linear frequency modulation coefficients in two distinct time intervals, this design achieves near-orthogonal symbol generation. Another method, discussed in [6, 8, 12, 28], uses loosely orthogonal chirps with varying modulation coefficients, overlapping them within a single time duration to increase the modulation order, as shown in

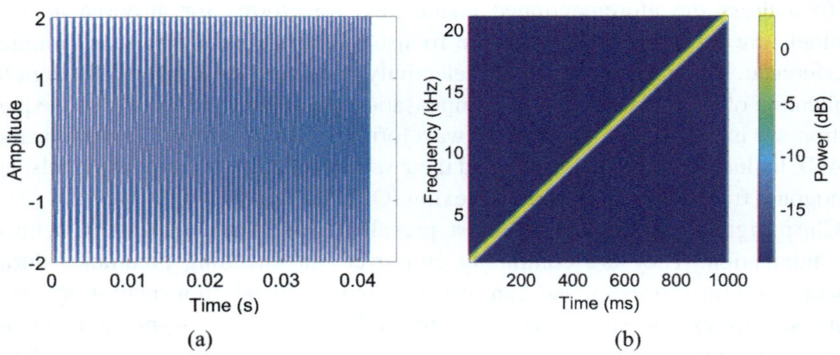

Fig. 2.8 Chirp signals in time and time-frequency domain. (**a**) Chirp signal in time domain. (**b**) Chirp signal in time-frequency domain

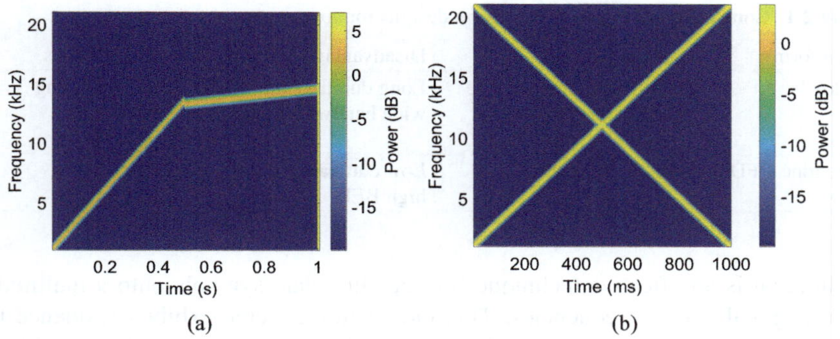

Fig. 2.9 Chirp variants. (**a**) A variant of chirp used to constitute orthogonal symbols. (**b**) Loose orthogonal chirps

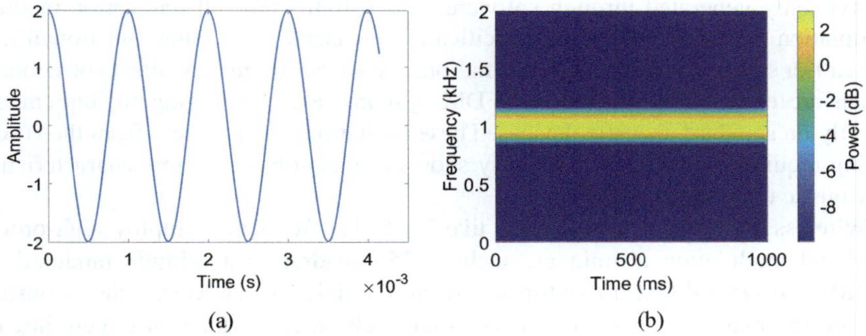

Fig. 2.10 Pure tone signals in time and time-frequency domain. (**a**) Pure tone signal in time domain. (**b**) Pure tone signal in time-frequency domain

Fig. 2.9b. While this approach introduces self-interference due to the approximate orthogonality, it proves effective in improving data rates.

A pure tone signal is another widely used waveform. It is expressed as

$$s(t) = a \cos(2\pi f t + \phi), \tag{2.14}$$

where a is the amplitude, and f and ϕ are the frequency and the initial phase. While pure tone signals offer simplicity in both encoding and decoding processes, they are less effective than chirp signals in mitigating noise and multipath fading. Figure 2.10 illustrates the representation of pure tone signals in both the time and frequency domains. Empirical studies (detailed in Chap. 4) reveal that the amplitude and phase of acoustic signals are prone to variations. Consequently, information is conveyed by modulating the carrier frequency.

An OFDM waveform, as its name implies, refers to an acoustic waveform modulated using orthogonal frequency division multiplexing (OFDM). OFDM

Table 2.1 Comparison of different waveform designs for communication

Waveform	Advantage	Disadvantage	Suitable applications
Chirp	Resilient to interference and Doppler effect	Long duration and wide bandwidth	Low throughput, long distance, low BER communication
Pure tone OFDM	High throughput	Low data rate and high BER	Short-range communication

modulation is an effective technique for encoding data symbols onto a multitude of orthogonal carrier frequencies. This modulation scheme exhibits resilience to synchronization errors and offers the potential for increased throughput within a given bandwidth [13, 21]. In wireless communication systems, OFDM modulation is typically implemented using specialized hardware to achieve near real-time performance. However, in the context of acoustic communication, OFDM signals are typically generated through software, which limits the full realization of this modulation scheme's potential. Specifically, advanced processing functionalities like carrier sensing, channel state estimation, and carrier frequency offset correction, which are essential in RF-based OFDM systems, are challenging to implement directly on standard acoustic devices. These limitations mainly stem from the strict timing requirements and the relatively short channel coherence time characteristic of acoustic environments [30].

Wireless communication systems like Wi-Fi [1] frequently employ high-order baseband modulation techniques, such as 256-quadrature amplitude modulation (QAM), within OFDM to optimize spectral efficiency. However, the acoustic domain presents challenges that render such high-order modulation largely infeasible due to significant channel impairments, including multipath propagation, Doppler effects, and frequency-selective fading. Furthermore, phase-based modulation schemes, which are common in RF systems, are rarely used in airborne acoustic communication because of the inherent instability of phase in acoustic channels. Instead, binary amplitude shift keying (ASK) is more typically adopted in acoustic OFDM implementations [33, 45]. As a result, OFDM in AAC functions in a multi-carrier Frequency Shift Keying (FSK) system, where information is primarily encoded using frequency and amplitude variations rather than phase.

Table 2.1 provides a summary of the benefits and drawbacks of various waveforms used in AAC, along with their typical applications.

2.3 Waveforms for Acoustic Sensing

Acoustic sensing leverages the propagation attributes of acoustic signals, including amplitude variations, phase shifts, Doppler frequency shifts, and spectral alterations, to identify and assess target objects or their movements. Consequently, the design of waveforms for acoustic sensing diverges significantly from those employed in

AAC. Typically, effective acoustic sensing necessitates that the chosen propagation characteristic exhibits substantial changes in response to environmental variations, thereby maximizing detectability and improving resolution. This principle stands in contrast to acoustic communication, where waveform design emphasizes signal resilience and minimal distortion to guarantee dependable data transmission.

Chirp signals, pure tones, and OFDM waveforms continue to be extensively used in acoustic sensing applications. Furthermore, various spread spectrum (SS) signals are also employed due to their advantageous characteristics. Similar to orthogonal polyphase coding waveforms used in radar systems [23, 50], SS signals exhibit an autocorrelation function resembling a thumbtack, with nearzero cross-correlation. These attributes make SS signals highly detectable, resistant to multipath effects, and robust to noise, rendering them appropriate for precise sensing applications.

The synthesis of SS signals begins with the selection of an orthogonal code [14] to serve as the baseband signal. Common orthogonal codes include, but are not limited to, the GSM training sequence [17], Barker code [44], M-sequence [15], and Zadoff-Chu (ZC) sequence [24].

The GSM training sequence is designed to mitigate channel distortions and equalize propagation delays in communication systems. It consists of eight distinct sequence types, each comprising 26-bit patterns. The eight GSM training sequence includes [42]:

$$
\begin{aligned}
&[0, 0, 1, 0, 0, 1, 0, 1, 1, 1, 0, 0, 0, 0, 1, 0, 0, 0, 1, 0, 0, 1, 0, 1, 1, 1;\\
&0, 0, 1, 0, 1, 1, 0, 1, 1, 1, 0, 1, 1, 1, 1, 0, 0, 0, 1, 0, 1, 1, 0, 1, 1, 1;\\
&0, 1, 0, 0, 0, 0, 1, 1, 1, 0, 1, 1, 1, 0, 1, 0, 0, 1, 0, 0, 0, 0, 1, 1, 1, 0;\\
&0, 1, 0, 0, 0, 1, 1, 1, 1, 0, 1, 1, 0, 1, 0, 0, 0, 1, 0, 0, 0, 1, 1, 1, 1, 0;\\
&0, 0, 0, 1, 1, 0, 1, 0, 1, 1, 1, 0, 0, 1, 0, 0, 0, 0, 0, 1, 1, 0, 1, 0, 1, 1;\\
&0, 1, 0, 0, 1, 1, 1, 0, 1, 0, 1, 1, 0, 0, 0, 0, 0, 1, 0, 0, 1, 1, 1, 0, 1, 0;\\
&1, 0, 1, 0, 0, 1, 1, 1, 1, 1, 0, 1, 1, 0, 0, 0, 1, 0, 1, 0, 0, 1, 1, 1, 1, 1;\\
&1, 1, 1, 0, 1, 1, 1, 1, 0, 0, 0, 1, 0, 0, 1, 0, 1, 1, 1, 0, 1, 1, 1, 1, 0, 0].
\end{aligned}
\tag{2.15}
$$

Each row in Eq. (2.15) represents a distinct type of GSM training sequence.

A Barker code is a commonly used sequence for intra-pulse biphase modulation in pulse compression radar systems [5, 37]. Barker codes are defined for lengths ranging from 2 to 13 bits:

$$
\begin{array}{lll}
\text{Barker} & \text{Code 2} & 10\\
\text{Barker} & \text{Code 3} & 110\\
\text{Barker} & \text{Code 4} & 1011\\
\text{Barker} & \text{Code 5} & 11101\\
\text{Barker} & \text{Code 7} & 1110010\\
\text{Barker} & \text{Code 11} & 11100010010\\
\text{Barker} & \text{Code 13} & 1111100110101.
\end{array}
\tag{2.16}
$$

The maximal-length sequence (M-sequence) is extensively utilized in 5G systems for synchronization [22]. It is generated using a linear feedback shift register (LFSR), which produces a deterministic yet pseudo-random bit sequence with desirable autocorrelation properties. Given an n-stage LFSR, the bit sequence at the kth time instant is mathematically represented as [22]

$$s_k = (s_{k,0}, s_{k,1}, s_{k,2}, ..., s_{k,n-1}), \tag{2.17}$$

where $s_{k,i}$ is the ith bit at kth moment. The bit sequence at $(k+1)$th moment can be generated using the follow formula [22]:

$$s_{k+1} = s_k \oplus (C_0 \cdot s_{k,0} + C_1 \cdot s_{k,1} + C_2 \cdot s_{k,2} +, ..., + C_{n-1} \cdot s_{k,n-1}), \tag{2.18}$$

where $C_k \in [0, 1]$ denotes the coefficient and $\oplus$ denotes XOR operation. The presence of C_k indicates the participant of the kth bit in deducing the bit sequence for the next state.

Yet another option is ZC sequence [3, 19, 20, 26]. The ZC sequence is a fundamental component of spread spectrum techniques in modern cellular systems, including LTE and 5G NR. It possesses several advantageous properties, such as constant amplitude, zero cyclic autocorrelation, and ideal cross-correlation characteristics, making it highly effective for synchronization and channel estimation. The given key parameters of a ZC sequence include the length $\mathcal{N}_{\mathrm{ZC}}$ and the the root index $q \in [1, \mathcal{N}_{zc}]$. Given these two parameters, a ZC sequence is defined as:

$$s_q[n] = \exp\left[-j\pi q \frac{n(n+1)}{\mathcal{N}_{\mathrm{ZC}}}\right], \tag{2.19}$$

where $\mathcal{N}_{\mathrm{ZC}}$ is a prime number. Figure 2.11 shows the autocorrelation performance among different orthogonal sequences. In the figure, we use the first row in Eq. (2.15) for GSM, code 13 for Barker code, a polynomial $x^5 + x^4 + x^3 + x^2 + 1$ for the M-sequence, and $q = 7, \mathcal{N}_{\mathrm{ZC}} = 37$ for the ZC. Results indicate that ZC sequence achieves the best autocorrelation performance: There is only one sharp peak at zero delay.

Table 2.2 summarizes the key advantages and disadvantages as well as the use cases of different waveforms for acoustic sensing. A quantitative comparison among different waveforms for acoustic sensing can be found in [30].

Synthesizing a pass-band signal using the previously mentioned orthogonal sequences involves several steps. Initially, system parameters such as the signal bandwidth B, the sampling rate f_s, the carrier frequency f_c, and the duration T of the sensing waveform must be defined. Subsequently, an orthogonal code X_{OC} of length BT is generated. If the length does not precisely equal BT, it is preferable to slightly adjust the duration T to preserve the autocorrelation properties of the selected code. The resulting code is then up-sampled at an interpolation factor of f_s/B to produce the baseband signal $s_{\mathrm{base}}(t)$. This interpolation can also be

Fig. 2.11 Autocorrelation performance of different sequences

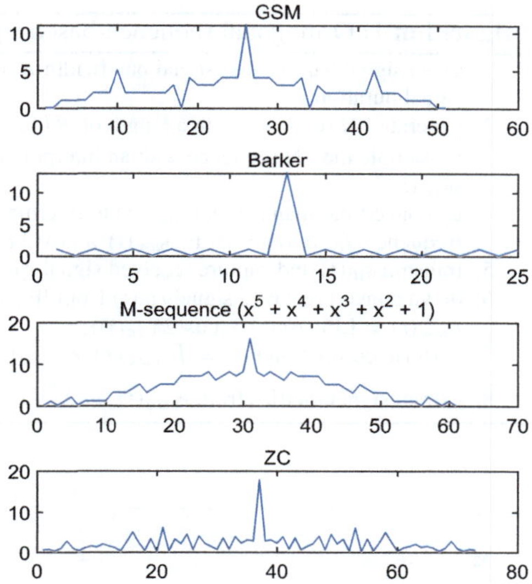

Table 2.2 Comparison of different waveform designs for sensing

Waveform	Advantage	Disadvantage	Suitable applications
Chirp	Resilient to interference and Doppler effect	Long duration and wide bandwidth less sharp correlation	Synchronization
Pure tone	Doppler sensitive and can be inaudible	Low SNR easily interfered	Short-range sensing
Orthogonal code enabled waveform	High sensitivity and high multipath resolution	Low SNR easily interfered wide bandwidth audibility issue	Short-range sensing

performed in the frequency domain by appending zeros to the end of the signals. Finally, the baseband signal is up-converted into an intermediate frequency (IF) signal $s_{IF}(t)$ by a frequency shift operation, $s_{IF}(t) = s_{base}(t)\cos(2\pi f_c t)$. $s_{IF}(t)$ is then ready to be transmitted through speakers. At the receiver end, the above process is reversed. Given a received IF signal is $r_{IF}(t)$, the first step is to down-convert the signals into baseband $r_{base}(t) = \text{LPF}\{r_{IF}(t)\cos(2\pi f_c t)\}$. A correlation operation between $r_{base}(t)$ and $s_{base}(t)$ can then be performed for a precise synchronization. Subsequently, various signal characteristics can be extracted from $r_{base}(t)$ for sensing, including phase, spectrum, etc. The complete waveform synthesis and processing steps are outlined in Algorithm 1.

Algorithm 1: Orthogonal sequence-based signal generation process

1: given signal parameters: signal bandwidth B, central frequency f_c, sampling rate f_s, signal duration T;
2: generate PN sequence with a length of BT;
3: up-sample the PN sequence with an interpolation rate of $\frac{f_s}{B}$ to get the baseband signal $s_{\text{base}}(t)$;
4: up convert baseband signal $s_{\text{base}}(t)$ to intermediate frequency band $s_{\text{IF}}(t)$ with central frequency f_c, $s_{\text{IF}}(t) = \text{BPF}\{s_{\text{base}}(t) \times \cos(2\pi f_c t)\}$;
5: transmit $s_{\text{IF}}(t)$ and capture received signal $r_{\text{IF}}(t)$;
6: down convert received signal $r_{\text{IF}}(t)$ from IF band to baseband $r_{\text{base}}(t)$, $r_{\text{base}}(t) = \text{LPF}\{r_{\text{IF}}(t) \times \cos(2\pi f_c t)\}$;
7: perform correlation $r(t) = \int_\tau s_{\text{base}}(\tau) \times r_{\text{base}}(t-\tau) d\tau$;
8: extract signal metrics from $r_{\text{base}}(t)$.

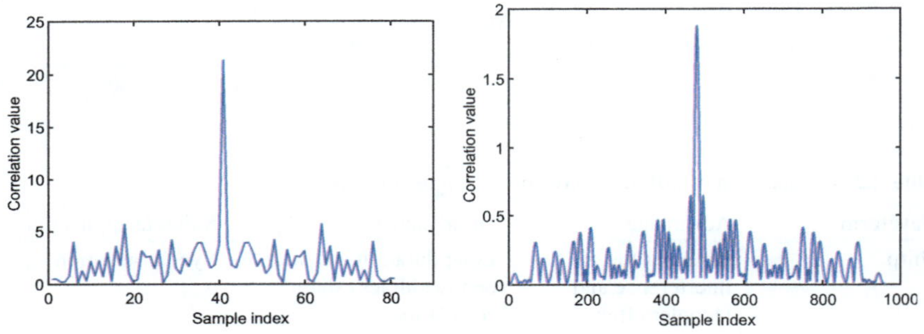

Fig. 2.12 Time domain autocorrelation performance of ZC sequence and baseband ZC signal. (**a**) Autocorrelation performance of ZC sequence. (**b**) Autocorrelation performance of baseband ZC signal

An Example

The process of generating a desired orthogonal code for sensing can be illustrated using the complex ZC sequence as an example.

Consider a sensing application operating at a frame rate of 100 Hz, or with a signal duration of $T = 10$ms. At a sampling frequency of 48 kHz, this corresponds to 480 samples per frame. Suppose the available bandwidth is $B = 4$ kHz, and the carrier frequency is set to $f_c = 20$ kHz, ensuring that the IF signal spans the 18–22 kHz range, to remain *inaudible to humans*.

With a bandwidth of $B = 4$ kHz, the required length of the orthogonal code is determined by $B \times T = 4000 \times 0.01 = 40$. Since ZC sequences require a prime-length sequence, we select $\mathcal{N}_{\text{ZC}} = 41$ with a root index $q = 7$. The autocorrelation function of the generated ZC sequence, demonstrating its excellent correlation properties, is illustrated in Fig. 2.12a.

The subsequent stage entails transforming the ZC sequence into a baseband ZC signal. This transformation can be achieved either by interpolation in the time domain or, as an alternative, by zero padding in the frequency domain. For the sake of maintaining spectral consistency, we choose the latter method. Given a frame duration of 10 ms, the resulting baseband ZC signal is required to have 480 samples. Therefore, the original 41-sample ZC sequence is interpolated to become a 480-sample baseband signal, while preserving a 4 kHz bandwidth. Considering that the ZC sequence is inherently a dual-band signal, we apply the fast Fourier transform (FFT) to the original 41-sample ZC sequence and then divide the resulting frequency spectrum into two symmetrical parts. Specifically, the first 21 samples, which include the DC component, are placed at the first 21 indices of the frequency spectrum. The remaining samples are then positioned symmetrically at the highest frequency indices to ensure the integrity of the spectrum. The time-frequency representation of the resulting baseband ZC signal is illustrated in Fig. 2.13b, and its autocorrelation properties are depicted in Fig. 2.12b.

After interpolation, the baseband signal is up-converted into an IF signal following Step 4 in Algorithm 1. The IF signal can be generated by extracting its real component:

$$s_{\text{IF}}(t) = \text{Re}\{s_{\text{base}}(t)e^{j2\pi f_c t}\}.$$

Alternatively, the real and imaginary components of the baseband ZC signal can be separately up-converted as follows:

$$s_{\text{IF}}(t) = \text{Re}(s_{\text{base}}(t))\cos(2\pi f_c t) - \text{Im}(s_{\text{base}}(t))\sin(2\pi f_c t),$$

where $\text{Re}(\cdot)$ and $\text{Im}(\cdot)$ denote the real and imaginary components of the baseband ZC signal, respectively. The time-frequency analysis of the IF ZC signal is depicted in Fig. 2.14a. The final IF signal is then transmitted through a speaker.

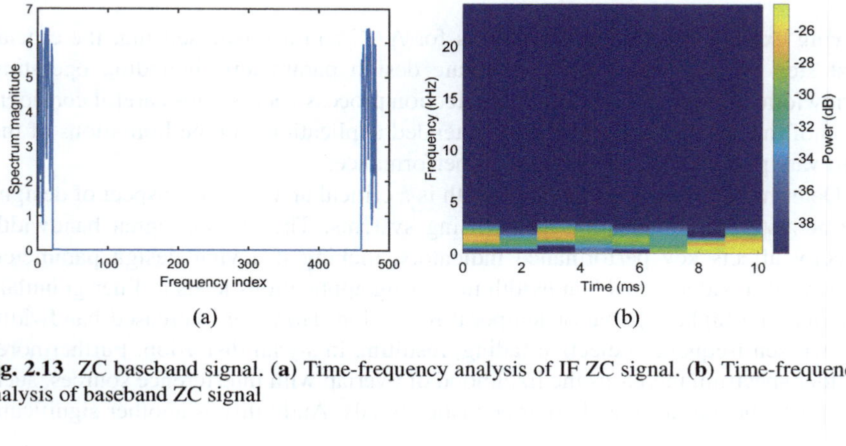

(a) (b)

Fig. 2.13 ZC baseband signal. (**a**) Time-frequency analysis of IF ZC signal. (**b**) Time-frequency analysis of baseband ZC signal

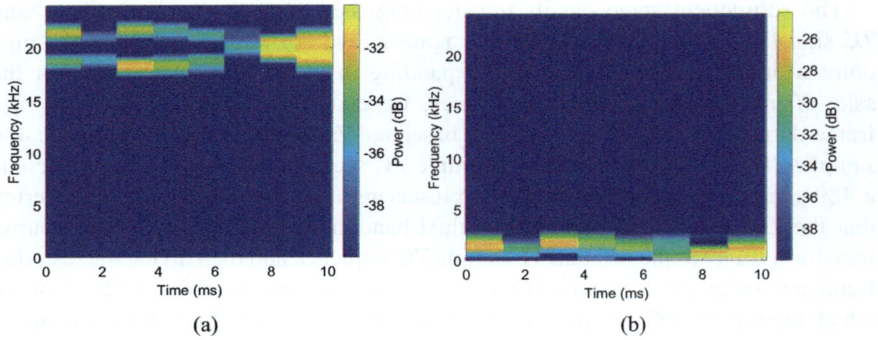

Fig. 2.14 Time-frequency analysis of ZC IF signal and received baseband signal. (**a**) Time-frequency analysis of IF ZC signal. (**b**) Time-frequency analysis of received baseband ZC signal

At the receiver end, the received IF signal $r_{\text{IF}}(t)$ undergoes down-conversion back to the baseband:

$$r_{\text{base}}(t) = \text{LPF}\{r_{\text{IF}}(t)\cos(2\pi f_c t)\},$$

where LPF removes high-frequency components introduced during mixing. The resulting time-frequency representation of the received baseband signal is shown in Fig. 2.14b. After low-pass filtering, the baseband signal is ready for feature extraction, including phase, spectrum, and time of flight (ToF) analysis for sensing applications.

2.4 Other Considerations

Having examined common waveforms for AAC and acoustic sensing, the crucial next step involves selecting appropriate design parameters, including operating bandwidth and signal duration. This selection process necessitates careful consideration of the specific demands of the intended application and the limitations of the hardware platform to achieve optimal performance.

Determining the operating bandwidth is a critical and complex aspect of designing acoustic communication and sensing systems. The chosen signal bandwidth directly affects key performance indicators, making it a vital design parameter. Generally, a wider signal bandwidth in sensing applications leads to finer granularity, enabling higher spatial or temporal resolution. However, increased bandwidth can worsen frequency-selective fading, resulting in signal distortion. Furthermore, a wider spectrum increases the likelihood of overlap with interference sources, such as environmental noise and other acoustic signals. Audibility is another significant consideration, particularly in airborne acoustic applications, where human hearing can be a limiting factor. Signals in lower frequency ranges (0–10 kHz) may produce audible artifacts, while higher frequencies (above 18 kHz) might be affected by limitations in the frequency response of speakers and microphones found in common devices. One potential approach to mitigate these issues is to use broadband signals with a white-noise-like characteristic and low signal magnitude [7].

The duration of the designed signal is another crucial parameter that demands careful consideration. It directly influences signal quality, robustness, and vulnerability to multipath interference, necessitating a delicate balance between these competing aspects. On one hand, a longer signal duration generally improves the signal-to-noise ratio (SNR). For instance, the processing gain of a chirp signal is proportional to the product of its bandwidth and duration. Thus, extending the signal duration enhances energy accumulation over time, leading to improved SNR and detection capabilities. Additionally, a longer duration facilitates better noise averaging, enhancing overall signal integrity in noisy conditions. Conversely, excessively long signal durations can be detrimental. As the signal duration increases, the latter part of the signal becomes more prone to multipath reflections and delayed echoes, which can distort the received signals and reduce decoding accuracy. In real-time applications, a longer signal duration increases processing latency, which can be problematic for tasks requiring low latency, such as interactive sensing or real-time tracking.

In conclusion, by judiciously choosing suitable signal bandwidth and duration for waveforms, designers can optimize signal quality while minimizing distortion and interference, thereby ensuring robust performance across diverse environments.

References

1. *256 QAM - 802.11ac*. https://wirelessonthego.postach.io/post/256-qam-802-11ac (2015)
2. *Acoustic Resonance*. https://cmtext.indiana.edu/acoustics/chapter1_resonance (2023)
3. J. Andrews. *A Primer on Zadoff Chu Sequences*. Nov. 2022.
4. B. Ballot. *Akustische Versuche auf der Niederla¨ndischen Eisenbahn, nebst gelegentlichen Bemerkungen zur Theorie des Hrn. Prof. Doppler* (1845)
5. *Barker Code*. https://helpfiles.keysight.com/csg/n7620b/Content/Main/Barker_Code.htm (2021)
6. C. Cai et al., Asynchronous acoustic localization and tracking for mobile targets. IEEE Int. Things J. **7**, 830–845 (2019)
7. C. Cai et al., AcuTe: Acoustic thermometer empowered by a single smartphone, in *Proceedings of the 18th ACM SenSys* (2020), pp. 28–41
8. C. Cai et al., Boosting chirp signal based aerial acoustic communication under dynamic channel conditions. IEEE Trans. Mob. Comput. **21**, 1–11 (2021)
9. C. Cai et al., We hear your PACE: Passive acoustic localization of multiple walking pErsons, in *Proceedings of the ACM Ubicomp* (2021), pp. 1–23
10. C. Cai, R. Zheng, J. Luo, Ubiquitous acoustic sensing on commodity IoT devices: A survey. IEEE Commun. Surv. Tutor. **24**(1), 432–454 (2022)
11. C. Cai et al., Active acoustic sensing for "Hearing" temperature under acoustic interference. IEEE Trans. Mob. Comput. **22**(2), 661–673 (2023)

12. C. Cai et al., Reliable high throughput aerial acoustic communication for mobile network. IEEE Trans. Vehic. Technol. **73**(4), 5704–5716 (2024)
13. Y. Chapre et al., Received signal strength indicator and its analysis in a typical WLAN system (short paper), in *Proceedings of the 38th IEEE Conference on Local Computer Networks* (2013), pp. 304–307
14. F.R.K. Chung, J.A. Salehi, V.K. Wei, Optical orthogonal codes: design, analysis and applications. IEEE Trans. Inf. Theory **35**(3), 595–604 (1989)
15. M. Cohn, A. Lempel, On fast M-sequence transforms (Corresp.). IEEE Trans. Inf. Theory **23**(1), 135–137 (1977)
16. *Doppler Effect*. https://www.britannica.com/science/Dopplereffect (2021)
17. C. Drane, M. Macnaughtan, C. Scott, Positioning GSM telephones. IEEE Commun. Mag. **36**(4), 46–54 (1998)
18. A. Ferlini et al., EarGate: Gait-based user identification with in-ear microphones, in *Proceedings of the 27th Annual International Conference on Mobile Computing and Networking* (2021), pp. 337–349
19. M.M.U. Gul, S. Lee, X. Ma, Robust synchronization for OFDM employing Zadoff-Chu sequence, in *2012 46th Annual Conference on Information Sciences and Systems (CISS)*. CISS (2012)
20. M.M.U. Gul, X. Ma, S. Lee, Timing and frequency synchronization for OFDM downlink transmissions using Zadoff-Chu sequences. IEEE Trans. Wirel. Commun. **14**(3), 1716–1729 (2015)
21. J. Heiskala, J. Terry, *OFDM Wireless LANs: A Theoretical and Practical Guide* (Sams, Carmel, 2001)
22. T. Helleseth, Maximal-length sequences, in *Encyclopedia of Cryptography and Security*, ed. by H.C.A. van Tilborg, S. Jajodia (Springer, Boston, 2011), pp. 763–766
23. H. Hu, B. Liu, Genetic algorithm for designing polyphase orthogonal code, in *2008 4th International Conference on Wireless Communications, Networking and Mobile Computing* (2008), pp. 1–4
24. M. Hua et al., Analysis of the frequency offset effect on Zadoff–Chu sequence timing performance. IEEE Trans. Commun. **62**(11), 4024–4039 (2014)
25. L. Huang et al., Acoustic resonances in non-Hermitian open systems. Nat. Rev. Phys. **6**, 1–17 (2023)
26. M. Hyder, K. Mahata, Zadoff–Chu sequence design for random access initial uplink synchronization in LTE-like systems. IEEE Trans. Wirel. Commun. **16**(1), 503–511 (2017)
27. R.S.H. Istepanian, M. Stojanovic, *Underwater Acoustic Digital Signal Processing and Communication Systems* (Springer, Berlin, 2002)
28. R. Jin et al., MotionBeep: enabling fitness game for collocated players with acoustic-enabled IoT devices. IEEE Int. Things J. **8**(13), 10755–10765 (2021)
29. S. Ka et al., Near-ultrasound communication for TV's 2Nd screen services, in *Proceedings of the 22Nd Annual International Conference on Mobile Computing and Networking*. MobiCom '16 (2016)
30. S. Ke et al., VSkin: Sensing touch gestures on surfaces of mobile devices using acoustic signals, in *Proceedings of the 24th ACM MobiCom* (2018), pp. 591–605
31. D. Kundu, S. Nandi, *Statistical Signal Processing*, vol. 83 (Springer, Berlin, 2021), pp. 844–890
32. P.M. Mariano, M. Spadini, Sources of finite speed temperature propagation. J. Non-Equilib. Thermodyn. **47**(2), 165–178 (2022)
33. R. Nandakumar et al., Dhwani: Secure peer-to-peer acoustic NFC, in *Proceedings of the ACM SIGCOMM 2013 Conference on SIGCOMM*. SIGCOMM '13 (2013)
34. A. Novák, P. Lotton, L. Simon, Synchronized swept-sine: theory, application, and implementation. J. Audio Eng. Soc. **63**, 786–798 (2015)
35. C. Peng et al., Beepbeep: A high accuracy acoustic ranging system using cots mobile devices, in *Proceedings of the 5th ACM SenSys* (2007), pp. 1–14

36. J. Prakash et al., EarSense: Earphones as a teeth activity sensor, in *Proceedings of the 26th ACM MobiCom* (2020), pp. 1–13
37. *Radar tutorial: Barker Codes*. https://www.radartutorial.eu/08.transmitters (2021)
38. A. Ross, G. Ostiguy, Propagation of the initial transient noise from an impacted plate. J. Sound Vib. **301**, 28–42 (2007)
39. N. Roy, H. Hassanieh, R.R. Choudhury, BackDoor: Making microphones hear inaudible sounds, in *Proceedings of the 15th ACM MobiSys* (2017), pp. 2–14
40. W. Ruan et al., AudioGest: Enabling fine-grained hand gesture detection by decoding echo signal, in *Proceedings of the 18th ACM UbiComp* (2016), pp. 474–485
41. S. Shen et al., Voice localization using nearby wall reflections, in *Proceedings of the 26th ACM MobiCom* (2020)
42. J. Shi, J. Huang, K. Liu, A training sequence parallel detection technology based on timeslot sliding window, in *The Proceedings of the Second International Conference on Communications, Signal Processing, and Systems*, ed. by B. Zhang et al. (Springer International Publishing, Cham, 2014), pp. 805–815
43. *The physics Classroom*. https://www.physicsclassroom.com/class/sound/Lesson-2/Intensity-and-the-Decibel-Scale (2021)
44. S. Wang, P. He, Research on low intercepting radar waveform based on LFM and barker code composite modulation, in *2018 International Conference on Sensor Networks and Signal Processing (SNSP)* (2018), pp. 297–301
45. Q. Wang et al., Messages behind the sound: Real-time hidden acoustic signal capture with smartphones, in *Proceedings of the 22Nd Annual International Conference on Mobile Computing and Networking*. MobiCom '16 (2016)
46. W. Wang et al., Symphony: Localizing multiple acoustic sources with a single microphone array, in *Proceedings of the 18th Conference on Embedded Networked Sensor Systems* (2020), pp. 82–94
47. W. Wang et al., Meta-speaker: Acoustic source projection by exploiting air nonlinearity, in *Proceedings of the 29th International Conference on Mobile Computing and Networking, ACM MobiCom'23* (ACM, Madrid, 2023)
48. Y. Yang et al., BioCase: Privacy protection via acoustic sensing of finger touches on smartphone case mini-structures, in *Proceedings of the 21st Annual International Conference on Mobile Systems, Applications and Services* (2023), pp. 397–409
49. Y. Zhang et al., Acoustic sensing and communication using metasurface, in *20th USENIX Symposium on Networked Systems Design and Implementation (NSDI 23)* (2023), pp. 1359–1374
50. X.-T. Zhao et al., Orthogonal polyphase code design for MIMO radar based on greedy algorithm, in *2021 International Conference on Electronic Information Engineering and Computer Science (EIECS)* (2021), pp. 119–122
51. B. Zhou et al., BatMapper: Acoustic sensing based indoor floor plan construction using smartphones, in *Proceedings of the 15th ACM MobiSys* (2017), pp. 42–55

Chapter 3
Basic Building Blocks for Acoustic Sensing

3.1 System Model

Let $x(t)$ be the input signal (e.g., a sound wave emitted by a source), $h(t, \tau)$ be the impulse response of the channel, and $y(t)$ be the output signal (the sound received after it has passed through the channel). When both the acoustic source and receiver are stationary and there is little change in the environment, the channel can be modeled as linear time invariant (LTI) and thus $h(t, \tau) = h(t - \tau)$. In contrast, when the relative positions of the source, receiver, and reflecting objects vary with time, the impulse response is a function of time and the time delay between the input and output. The system is referred to as a linear time-variant (LTV) system. In both cases, the output signal $y(t)$ is given by

$$y(t) = \int_{-\infty}^{\infty} x(\tau)h(t, \tau)\, d\tau + w(t),$$

where $w(t)$ is receiver noise or interference. The impulse response $h(t, \tau)$ characterizes the properties of the channel. In the case of acoustic propagation, this impulse response includes:

- Direct path: The signal traveling directly from the source to the receiver
- Reflections: Signals bouncing off walls, objects, etc.
- Attenuation and scattering: Energy losses due to the medium (air, water, etc.)

It is common to consider discrete-time and causal systems, where *causality* is defined as a system where the output depends on past and current inputs but not future inputs. Therefore,

$$y[n] = (x * h)[n] = \sum_{k=-\infty}^{n} x[k]h[n, k] + w(n),$$

R. Zheng, C. Cai, *Acoustic Sensing on Commodity Devices and its Applications*, Wireless Networks, https://doi.org/10.1007/978-3-031-96875-4_3

where $x[n]$ is the sampled input signal, $h[n]$ is the discrete impulse response, and $y[n]$ is the output signal after passing through the channel. Under the assumption of an LTI and finite impulse response (FIR) channel, we have

$$y[n] = (x * h)[n] = \sum_{k=0}^{L-1} x[n-k]h[k] + w(n), \tag{3.1}$$

where L is the number of taps, or the order of the FIR model, representing the duration of the channel's impulse response. In acoustic channels, L can be viewed at the number of non-negligible paths (including direct and multipaths).

3.2 Onset Detection

Onset detection is the process of identifying the precise moments in a signal where significant events or changes begin. In the context of active acoustic sensing, it refers to the determination of the presence and the start time of a particular reference signal. Onset detection is the cornerstone of acoustic communication and time-sensitive sensing applications.

To perform onset detection, the input signal is typically first divided into overlapping fixed length windows. Windowing is not strictly necessary but can reduce computation costs. For each window length of data, the binary hypothesis includes the null hypothesis (H_0), which assumes the absence of the reference signal, and the alternative hypothesis (H_1), which represents its presence. This can be achieved by performing cross-correlation between the windowed samples and the known *reference signal*. The waveform of the reference signal is carefully designed so that there exists a sharp peak in the correlation results when the reference signal is present; otherwise, the maximum correlation value is far below a predefined threshold (see illustration in Fig. 3.1). As a result, in the absence of multipath effects or strong interference, a simple threshold-based approach is sufficient to detect the presence of the reference signal, and the timestamp of the maximum peak is recorded as the onset of the signal.

In practice, the accuracy of threshold-based onset detection is degraded by three phenomena, namely, device heterogeneity, the *near-far* effect, and the multipath effects. Device heterogeneity refers to the fact speakers and microphones on different devices have different gains. A constant threshold may not work for all devices. The near-far effect, a term originated from wireless communication systems, describes the phenomenon where the signal power received at a base station is dominated by the signals from closer user devices due to signal attenuation over distance. Similarly, in acoustic systems, when a transmitter and a receiver are close, the resulting correlation peak at the receiver can readily exceed a predefined threshold, and hence, the reference signal is detected (Fig. 3.2a). However, when their distance becomes larger, the respective correlation peak may fall below the

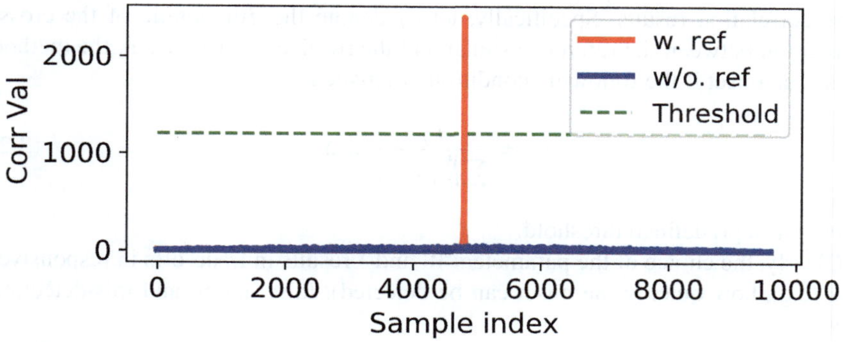

Fig. 3.1 Illustration of signal onset detection

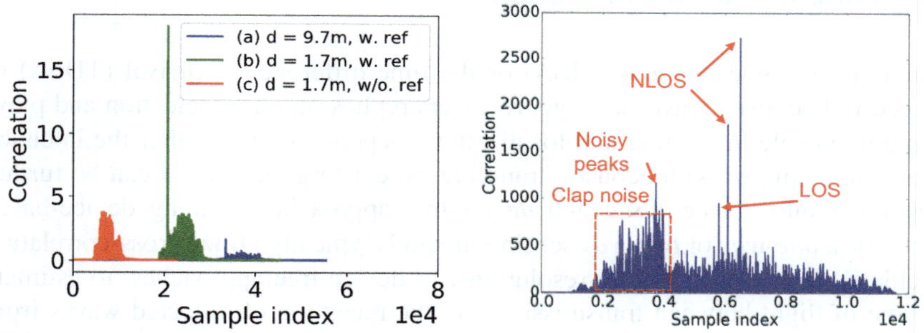

(a) Near-far effect. (b) Clap noise and multipath effect.

Fig. 3.2 (**a**) The correlation peak of distant samples is even weaker than the noisy peaks from closer ones, making it challenging to set an appropriate threshold for reliable onset detection. (**b**) Interference including clap noise and multipath effect can lead to incorrect onset estimation

threshold and in fact sometimes below the values caused by close-by interfering sources. In a multipath-rich environment, a receiver not only captures the line-of-sight (LoS) signal but also receives multiple delayed and attenuated copies. These delayed and attenuated copies, called none-line-of-sight (NLoS) signals, can add up constructively and thus surpass LoS signals in peak intensity. A naive thresholding approach or simply finding the largest peak will result in false onset detection. Figure 3.2 illustrates the effects of the latter two phenomena when the reference signal is a chirp signal. In Fig. 3.2a, hand claps prior to the transmission of the chirp signal result in noticeable peaks even after cross-correlation at the receiver. As shown in Fig. 3.2b, peaks due to NLOS paths are in fact larger than that from the LOS path.

In [1], a reliable preamble detection method is proposed to mitigate the effects of device diversity and the near-far problem. The main idea is to normalize the cross-correlation values by the sum intensity of the preceding W samples. Doing so effectively adapts the threshold dynamically based on the relative value of cross-correlation results. Specifically, let c_j denote the jth sample of the cross-correlation between the reference signal and the received signal. Then, the method detects an onset if the following condition is satisfied,

$$\tilde{c}_j = \frac{c_j}{\sum_{i=0}^{W} c_{j-i}} > \delta, \tag{3.2}$$

where δ is a predefined threshold.

Clearly, the choice of the parameters W and δ results in trade-offs in responsiveness (e.g., how quickly the onset can be detected), false alarm, and mis-detection rates.

3.3 Time-of-Flight Estimation

Estimating the time of arrival (ToA) or the time difference of arrival (TDoA) of the received acoustic waves at target receivers replies on onset detection and plays an important role in ranging and localization. Depending on whether the intended targets are equipped with acoustic transceivers, existing techniques can be further categorized into *device-based* and *device-free* approaches. Among device-based approaches, *one-way* or *two-way* sensing methods typically utilize cross-correlation to achieve sample-level timing resolutions. In device-free approaches, to estimate the time of flight between transmission and the reception of reflected waves from a target, phase information can also be used to achieve subsample-level resolutions because the transceiver and the receiver are typically colocated and share a common clock.

3.3.1 One-Way Sensing

One-way sensing generally refers to the sensing paradigm where time information is obtained through unidirectional transmissions. In this paradigm, transmitters and the receivers can reside on multiple spatially separated devices or a single device as shown in Fig. 3.3. In one-way sensing, tight synchronization between the transmitter and receiver is required so that a receiver knows precisely the onset of an acoustic transmission to estimate its flight time. Due to the limitation in commodity devices and the absence of shared clocks in distributed sensing applications, existing work achieves synchronization by exploiting another high-speed signal sources [10, 17]. Radio signals, such as WiFi, Bluetooth, and Zigbee, are suitable candidate sources because their propagation time is several orders of magnitude shorter than the flight time of acoustic waves within the maximum operational range. For ToA estimation, an acoustic signal and a synchronization signal are transmitted simultaneously. A receiver determines the ToA from the difference in the arrival time between the two

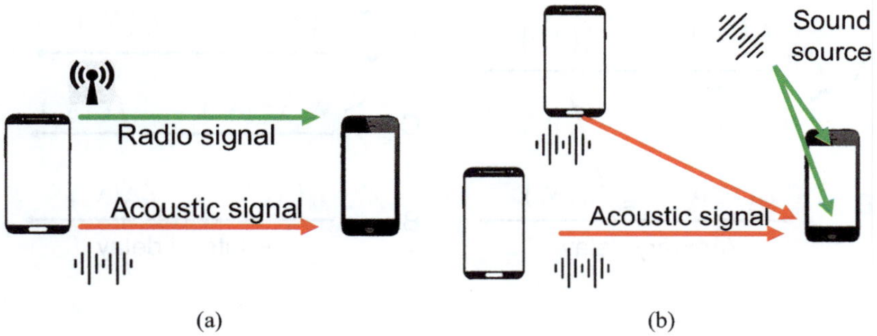

Fig. 3.3 One-way sensing paradigm. (**a**) ToA via one-way sensing. (**b**) TDoA via one-way sensing

signal sources. In this approach, timing information can be obtained without any coordination between the transmitter and the receiver.

TDoA refers to the difference in the time it takes for a signal from a single transmitter to arrive at two or more spatially separated receivers, or the time from multiple concurrent transmitters to a single receiver. For TDoA estimation, multiple transmitters or receivers need to be tightly synchronized. In some applications, transmitters or receivers colocate physically on a single device. In transmitter-synchronized systems [8, 9], acoustic transmissions are triggered concurrently, and they arrive at a receiver after different propagation delays as illustrated by the red arrows in Fig. 3.3b. TDoA is then obtained by estimating the onset of known reference signals. In receiver-synchronized systems, the TDoA of a single sound source is computed across multiple receiver channels on the same device (as indicated by the green arrows in Fig. 3.3b). This can be done by estimating the differences in the onsets by cross-correlating the received samples from different receive channels [11].

One-way sensing is a simple yet effective method for extracting timing information. However, its primary drawback is the requirement for tight synchronization, particularly in scenarios involving distributed devices. This synchronization requirement is often compromised by uncertain system delays, which can significantly degrade timing accuracy. Although these uncertainties can be mitigated through kernel-level implementations on general-purpose operating systems or customized hardware, the practicality and applicability of one-way sensing remain severely limited.

3.3.2 Two-Way Sensing

To eliminate the need for tight synchronization in one-way sensing, two-way sensing has been proposed, though it comes with increased hardware and processing

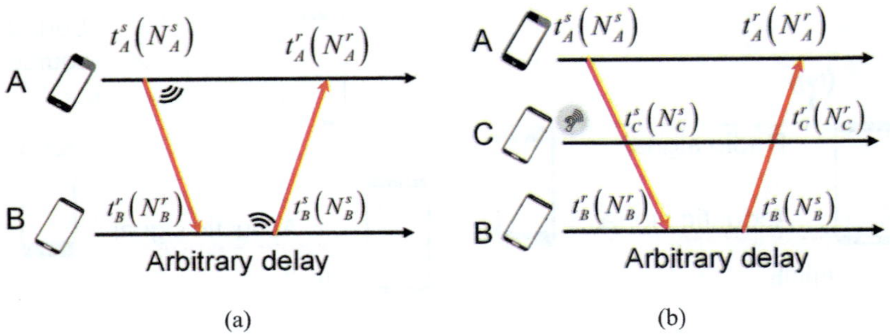

Fig. 3.4 Two-way sensing paradigm. (**a**) ToA via two-way sensing. (**b**) TDoA via two-way sensing

complexity. In two-way sensing, bidirectional acoustic transmissions are necessary. Consequently, each device should be equipped with both a speaker and a microphone. The steps to extract ToA information between two devices are shown in Fig. 3.4a. At time t_A^s, device A starts an acoustic transmission. Device B detects the acoustic signal at time t_B^r and starts another transmission at time t_B^s after an arbitrary delay. Device A detects the second transmission at time t_A^r. Under the assumption of reciprocal channels[1] with equal delay in both directions, the ToA from device A to device B can be derived as

$$\Delta_{\text{ToA}} = \frac{1}{2}\left(t_A^r - t_A^s\right) - \frac{1}{2}\left(t_B^s - t_B^r\right). \tag{3.3}$$

In the case of asynchronous transmitters, the TDoA between signals from transmitters A and B at receiver C with the consideration of the differences in transmission times (Fig. 3.4b) can be derived by

$$\Delta_{\text{TDoA}} = \frac{1}{2}\left(t_A^r - t_A^s\right) + \frac{1}{2}\left(t_B^r - t_B^s\right) - \left(t_C^r - t_C^s\right). \tag{3.4}$$

In practice, t_A^s, $t_A^r, t_B^r, t_B^s, t_C^r, t_C^s$ need to be time-stamped at the respective device's audio interfaces. In the presence of uncertain system delay, t_A^s, $t_A^r, t_B^r, t_B^s, t_C^r, t_C^s$ cannot be recorded precisely in user applications. To overcome such a challenge, two novel techniques have been proposed in [14]. First, in addition to transmissions from other devices, each transmitting device also records its own transmission through its on-board microphone. Second, sample counting in the audio buffer of a device is used to estimate the time elapsed between consecutive acoustic receptions. Specifically, as shown in Fig. 3.5a, upon transmitting a beacon

[1] A reciprocal channel refers to a communication channel whose channel characteristics are identical (or highly similar) in both transmission directions (up-link and down-link), assuming both directions use the same carrier frequency and experience the same propagation environment.

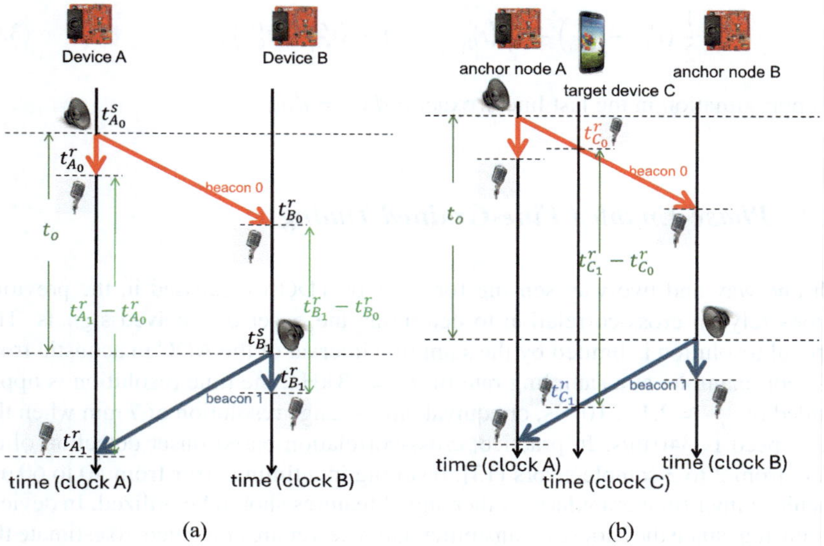

Fig. 3.5 Two-way sensing on commodity devices. Each device is assumed to be equipped with a speaker and a microphone. Timing differences between received signals are estimated via sample counting in the audio buffer. (**a**) ToA via two-way sensing. (**b**) TDoA via two-way sensing

signal, Device A (B) stores the received waveform in its audio buffer with timing $t^r_{A_0}$ ($t^r_{B_1}$). Accordingly, the received waveform from Device A (B)'s transmission is recorded at time $t^r_{B_0}$ ($t^r_{A_1}$). Using sample counting in the respective audio buffers, $t^r_{A_1} - t^r_{A_0}$ and $t^r_{B_1} - t^r_{B_0}$ can be estimated. Let the known distance from Device A (B)'s transmitter and receiver be d_{AA} (d_{BB}). Equation (3.3) can be rewritten as

$$\begin{aligned}\Delta_{\text{ToA}} &= \frac{1}{2}\left(t^r_{A_1} - t^s_{A_0}\right) - \frac{1}{2}\left(t^s_{B_1} - t^r_{B_0}\right)\\ &= \frac{1}{2}\left(t^r_{A_1} - t^r_{A_0} + \frac{d_{AA}}{c}\right) - \frac{1}{2}\left(t^r_{B_1} - t^r_{B_0} - \frac{d_{BB}}{c}\right)\\ &= \frac{1}{2}\left(t^r_{A_1} - t^r_{A_0}\right) - \frac{1}{2}\left(t^r_{B_1} - t^r_{B_0}\right) - \frac{1}{2c}\left(d_{BB} + d_{AA}\right),\end{aligned} \tag{3.5}$$

where c is the speed of sound in the medium. Using a similar argument, the TDoA estimation of the received signals from Devices A and B at a target device C (Fig. 3.5b) is expressed as [19]

$$\begin{aligned}\Delta_{\text{TDoA}} &= \frac{1}{2}\left(t^r_{A_1} - t^s_{A_0}\right) + \frac{1}{2}\left(t^s_{B_1} - t^r_{B_0}\right) - \left(t^r_{C_1} - t^r_{C_0}\right)\\ &= \frac{1}{2}\left(t^r_{A_1} - t^r_{A_0} + \frac{d_{AA}}{c}\right) + \frac{1}{2}\left(t^r_{B_1} - t^r_{B_0} - \frac{d_{BB}}{c}\right) - \left(t^r_{C_1} - t^r_{C_0}\right)\end{aligned}$$

$$\approx \frac{1}{2}\left(t^r_{A_1} - t^r_{A_0}\right) + \frac{1}{2}\left(t^r_{B_1} - t^r_{B_0}\right) - \left(t^r_{C_1} - t^r_{C_0}\right). \tag{3.6}$$

The approximation in the last line is exact if $d_{AA} = d_{BB}$.

3.3.3 *Phase-Enabled Fine-Grained Timing*

Both one-way and two-way sensing for TOA or TDOA discussed in the previous sections rely on cross-correlation to determine the onset of received signals. The temporal resolution is limited by the sampling interval of the ADC in acoustic front ends. For example, at a sampling rate of $f_s = 48$ kHz, the time resolution is upper bounded by $\frac{1}{f_s} \approx 2.1 \times 10^{-5}$ s, or equivalently a range resolution of 7 mm when the sound speed is 340 m/s. In practice, cross-correlation-based onset detection often suffers from 2 to 3 sample errors [13], resulting in a timing error from 40 to 60 μs. To achieve finer time granularity, other signal features should be utilized. In device-free sensing, since the acoustic transmitter and receiver are colocated, to estimate the time of flight of the reflected waves from interested targets, phase information can be exploited. Note that phase-enabled timing estimation techniques can be applied to device-based timing estimation as well. However, in systems where the transmitters and receivers are not colocated on the same devices, error sources such as carrier frequency offsets (CFO) [12] and sampling frequency offsets (SFO) can adversely affect the accuracy of such methods. CFO is the difference between the intended (ideal) carrier frequency and the actual carrier frequency generated or received due to oscillator inaccuracies, Doppler shifts, or hardware imperfections. SFO refers to the difference or mismatch between the intended (ideal) sampling rate and the actual sampling rate used in analog-to-digital (ADC) or digital-to-analog (DAC) conversions due to oscillator inaccuracies. More discussions on CFO and SFO and mechanisms to mitigate them can be found in Chap. 4.

The relationship between phase changes and time of flight is waveform dependent. We next present the basic approaches to extract phases from two waveforms discussed in Sect. 2.3, namely, pure tone signals and chirp signals.

Pure Tone Signals

The phase of pure tone signals can be extracted by a coherent receiver whose structure is shown in Fig. 3.6. In a coherent receiver, two identical copies of an input signal are multiplied by $\cos(2\pi f t)$ and its $\frac{\pi}{2}$ phase shifted version $-\sin(2\pi f t)$, respectively. After a low-pass filter, the in-phase (I) and quadrature-phase (Q) components can be obtained. The absolute phase ϕ in $s(t) = \cos(2\pi f t + \phi)$ is calculated by $\tan^{-1}(Q/I)$. Phase changes can be determined by subtracting consecutive absolute phase as illustrated in Fig. 3.7a. Consider a pure tone oscillating at $f_c = 20$ kHz sampled at $f_s = 48$ kHz. One sample interval corresponds to a phase change of $\frac{f_c}{f_s} \times 2\pi = \frac{5}{6}\pi$. As long as the detectable phase change is below $\frac{5}{6}\pi$, one can readily achieve subsample time resolution. Therefore, phase-based approaches

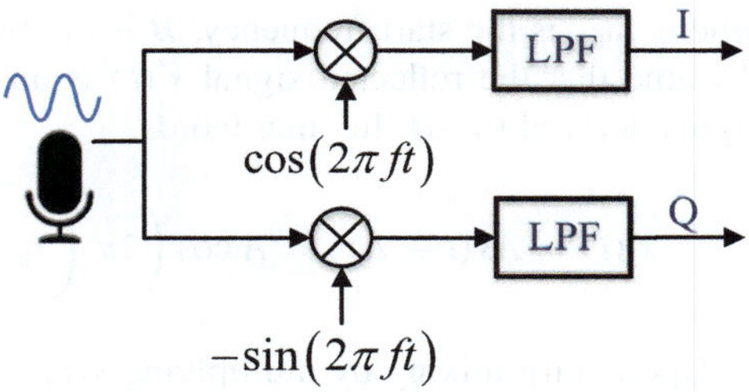

Fig. 3.6 Structure of a coherent receiver. LPF in the figure stands for low-pass filter

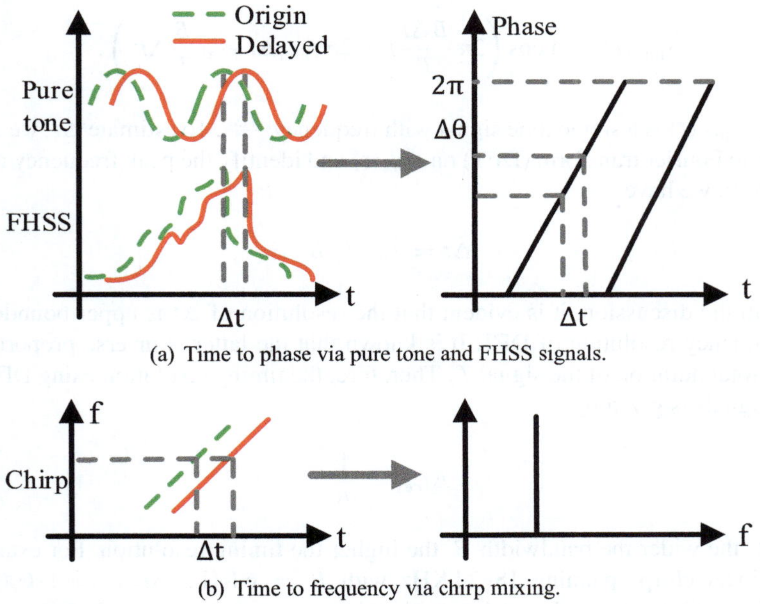

Fig. 3.7 **(a)** Temporal characteristics are converted into phase in pure tone and FHSS signal based system. **(b)** Chirp mixing transforms time measurements into frequency information

generally outperform cross-correlation-based ones in time granularity. However, a major shortcoming of pure tone signals is their vulnerability to background noise and multipath effects.

Chirp Signals
Extracting phases from chirp signals is achieved through chirp mixing, which converts time information into frequency features as illustrated in Fig. 3.7b. Let the transmitted linear chirp signal be

$$s(t) = \cos\left(2\pi\left(f_{\min}t + \frac{B}{2T}t^2\right)\right),$$

where $f_{\min}$ is the start frequency, B is the bandwidth, and T is the chirp duration. Assume that the reflected signal $s'(t)$ is an attenuated version of the transmitted signal delayed by Δt. In other words,

$$s'(t) = As(t - \Delta t) = A \cos\left(2\pi\left(f_{\min}(t - \Delta t) + \frac{B}{2T}(t - \Delta t)^2\right)\right).$$

Upon chirp mixing by multiplying $s(t) \times s'(t)$ and passing the resulting signal through a low-pass filter, we obtain

$$s_{\text{mix}}(t) = A \cos\left(2\pi \frac{B\Delta t}{T} t + 2\pi f_{\min}\Delta t - \pi \frac{B}{T}\Delta t^2\right).$$

Clearly, $s_{\text{mix}}(t)$ is a single tone signal with frequency $\frac{B\Delta t}{T}$. To estimate Δt, we apply a discrete Fourier transform (DFT) on $s_{\text{mix}}(t)$ and identify the peak frequency as f_p. Therefore, we have

$$\Delta t = f_p \cdot T/B.$$

From the discussion, it is evident that the resolution of Δt is upper bounded by the frequency resolution of DFT. It is known that the latter is inverse proportional by the total duration of the signal T. Therefore, the timing resolution using DFT for chirp signals is given by

$$\Delta t_{res} = \frac{1}{B}.$$

Clearly, the wider the bandwidth B, the higher the timing resolution. For example, for a linear chirp spanning 18–24 KHz with $B = 6$ KHz, $\Delta t_{res} = 1/6000 = 0.1667$ ms. Comparing this value with what can achieve through CC in time domain (in the order of tens of μS), this resolution is unsatisfactory. To further improve the timing resolution, phase information in the mixed signal can be utilized. Note in the frequency bin $f_p = \frac{B\Delta t}{T}$, the phase of the mixed signal $\phi(t) = 2\pi\left(f_{\min}\Delta t - \frac{B}{2T}\Delta t^2\right)$. Therefore, a time difference of 10 μs corresponds to a phase difference of 0.18 radian when $f_{\min} = 18$ kHz, significant enough to be detected. The detailed steps of phase-based timing estimation are summarized in Algorithm 2.

Phase Ambiguity

Phase-enabled fine-grained timing suffers from *phase ambiguity*, defined as indistinguishability in determining the exact phase of a periodic signal or waveform. It arises from the fact that the phase information of a signal is known only modulo 2π. For example, with a pure tone signal of frequency f and a phase difference of ϕ, the time difference can be $\left(k + \frac{\phi}{2\pi}\right) \cdot \frac{1}{f}, \forall k \in \mathbb{N}$. To help resolving phase ambiguity,

Algorithm 2: Estimate time delay (Δt) using chirp signal

1: **Input:** $s(t) = \cos\left(2\pi\left(f_{\min}t + \frac{B}{2T}t^2\right)\right)$ and delayed chirp signal $s'(t) = As(t - \Delta t)$
2: **Output:** Estimated time delay Δt
3: Mix the chirp signal with its delayed version:

$$x(t) = s(t) \cdot s'(t)$$

4: Apply a low-pass filter to $x(t)$ to remove high-frequency components:

$$x_{\text{LPF}}(t) = \text{LPF}(x(t))$$

5: Compute the Discrete Fourier Transform (DFT) of $x_{\text{LPT}}(t)$:

$$X(f) = \text{DFT}\{x_{\text{LPF}}(t)\}$$

6: Select the frequency bin f_p with the maximum energy in $X(f)$
7: Estimate the phase of the selected frequency bin:

$$\phi = \angle X(f_p)$$

8: Compute the time delay Δt using the phase and selected frequency:

$$\Delta t = \frac{\phi}{2\pi f_{\min}}$$

9: **return** Δt

multiple tones can be used. Consider two measured phase changes ϕ_1 and ϕ_2 for two tones of frequencies f_1 and f_2, respectively. The time difference then satisfies

$$\left(k + \frac{\phi_1 - \phi_2}{2\pi}\right) \cdot \frac{1}{|f_1 - f_2|}, \forall k \in \mathbb{N}.$$

Note that f_1 and f_2 should be chosen to have small differences to ensure a small "beat" frequency to be effective. Using chirp signals in phase-enabled timing estimation suffers from similar phase ambiguity. However, it is less pronounced since the procedure first performs frequency bin selection, which limits the uncertainty in Δt to be within Δt_{res}. When tracking phases over consecutive chirps, phase unwrapping is a technique to allow the estimation of continuous phase changes beyond the range of 2π [6].

3.4 Angle of Arrival Estimation

Determining the angle of arrival (AoA) is instrumental in localizing and tracking acoustic devices and interested targets. Various techniques for estimating the angle of arrival (AoA) of sound sources such as human speakers using microphone arrays have been proposed in the literature. These include time domain methods, such as TDoA-based AoA estimation [4], and spatial filtering methods, such as delay-and-sum beamforming (DSB), minimum variance distortionless response (MVDR), and MUltiple SIgnal Classification (MUSIC) [18], etc. These methods can be directly applied to the AoA estimation of active acoustic sources. However, the latter is inherently simpler because, unlike in passive sensing with general sound sources, active sensing benefits from prior knowledge of both the number of sources and the transmitted signal waveform.

For simplicity, we consider a linear array of N microphones spaced evenly along a straight line, called *uniform linear array (ULA)* (Fig. 3.8). Let the spacing between two adjacent microphones be d. The signal from a far-field source arrives at an angle θ relative to the end-fire direction[2] of the array.

3.4.1 TDoA-Based AoA Estimation

TDoA-based AoA estimation allows the determination of the direction of sound sources using the differences in arrival times of the signal at spatially separated microphones. The TDoA values are computed by measuring the relative delays between signal receptions. In a linear microphone array, the TDoA τ_{ij} between

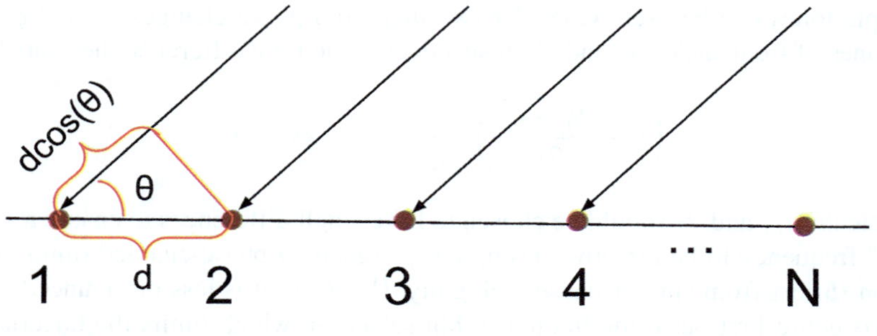

Fig. 3.8 Array geometry of a linear microphone array

[2] The end-fire direction of a ULA is defined as the direction along (parallel to) the axis of the antenna elements.

microphones i and j is given by

$$\tau_{ij} = \frac{(j-i)d\cos(\theta)}{c},$$

where c is the speed of sound in air. The TDoAs between microphone pairs can be estimated using timing estimation techniques discussed in Sect. 3.3. To determine the AoA, we can solve an optimization problem that minimizes the error between the measured TDoAs and the model-based TDoAs as a function of θ. The objective function is defined as

$$E(\theta) = \sum_{i=1}^{N-1}\sum_{j=i+1}^{N}\left(\hat{\tau}_{ij} - \frac{(j-i)d\cos(\theta)}{c}\right)^2,$$

where $\hat{\tau}_{ij}$ represents the measured TDoAs. The goal is to minimize this error function to estimate the AoA θ:

$$\hat{\theta} = \arg\min_{\theta} E(\theta). \tag{3.7}$$

The least-square estimator in Eq. (3.7) is the optimal maximum likelihood estimator under the assumption that the noises in $\hat{\tau}_{ij}, \forall i, j$ are independently and identically distributed zero-mean Gaussian variables. The above formulation can be easily extended to nonlinear microphone arrays by expressing TDoA as a function of the corresponding array geometry.

The accuracy of TDoA-based AoA is highly depended on the accuracy of timing estimation and array geometry. Due to the short distances among the microphones, phase-based methods are typically employed. Increasing the dimension of the array (i.e., aperture) and the number of microphones can improve AoA estimation resolution. Specifically, the angle resolution of a ULA is inversely proportional to N when the microphone is separately by half of a wavelength. Other sources of errors in AoA include multipath and interference from other sources.

3.4.2 *Spatial Filtering*

Beamforming is a spatial filtering technique to decompose signals coming from different directions. It is done by computing the (complex) weighted sum of the signals received at each microphone. When the complex weights are chosen appropriately, one can selectively extract or suppress signals from particular directions. AoA estimation can be accomplished by searching for the angles where the resulting beamformed signals have the largest amplitudes. Spatial filtering techniques can be broadly categorized into nonadaptive and adaptive spatial filtering. In nonadaptive spatial filtering, the beamforming weights (filter parameters) are predetermined and fixed for each spatial direction and are independent of the incoming signal and interference. Adaptive spatial filtering, on the other hand, adjusts the weights based on the characteristics of the incoming signal, aiming to optimize certain objective functions.

Delay-and-sum beamformer (DSB) is a form of nonadaptive spatial filtering. Denote by $y_m(t)$ the received signal at microphone $m, m = 0, 2, \ldots, N-1$. DSB calculates the uniformly weighted sum of delayed signals from the array. Specifically, for AoA θ',

$$\hat{y}_{\theta'}(t) = \sum_{m=1}^{N} y_m\left(t - \frac{md\cos(\theta')}{c}\right). \tag{3.8}$$

For a monochromatic source signal $x(t) = \cos(2\pi f t)$ of frequency f from angle θ, we have

$$y_m(t) = \Re\left(x(t)e^{\frac{2\pi md\cos\theta}{\lambda}}\right) + n_m(t),$$

where $\lambda = \frac{c}{f}$ is the wavelength and $n_m(t)$ is the received noise at microphone m.

Let $x_m(t) = x(t)e^{\frac{2\pi md\cos\theta}{\lambda}}$. For simplicity, expressing all signals in the complex domain, Eq. (3.8) can be rewritten as

$$\begin{aligned}\hat{y}_{\theta'}(t) &= \sum_{m=1}^{N}(x_m(t) + n_m(t))e^{-\frac{2\pi md(\cos\theta')}{\lambda}}\\ &= \mathrm{DFT}_u\{(x_m(t) + n_m(t)\}.\end{aligned} \tag{3.9}$$

Here, the spatial frequency in DFT satisfies $u = -\frac{d\cos\theta'}{\lambda}$. Therefore, in ULA, DSB to a direction θ' is equivalent to compute the DFT at spatial frequency u. DFT allows faster computation of DSB at various spatial angles. Additionally, it provides some insights to the performance of DSB-based AoA estimation. First, since the spatial frequency $u = -\frac{d\cos\theta'}{\lambda}$, the angle resolution of URA is not uniform in the angle domain. The angle resolution of DSB can be defined as its main lobe width. Thus, it is highest at boresight ($\theta = 90$) and decreases as θ decreases. Second, since the spatial frequency resolution of DFT is inverse proportional to N, an array with more elements has better angle resolution. Third, applying a window function, such as Hamming or Hanning windows [15], spatially (or equivalently, applying nonuniform weights) to the received signals across microphones can effectively reduce spectral leakage and enhance interference suppression. However, this improvement comes at the expense of reduced peak sharpness in the resulting array pattern and lower spatial resolution.

The MVDR beamformer, also known as the Capon beamformer, is an adaptive spatial filtering technique designed to minimize interference and noise power while preserving the desired signal's direction. Given a ULA with spacing d between neighboring microphones, the *steering vector* $\mathbf{a}(\theta)$ for a plane wave arriving at an angle θ is expressed as

$$\mathbf{a}(\theta) = \begin{bmatrix} 1 \\ e^{-j\frac{2\pi d}{\lambda}\cos(\theta)} \\ e^{-j\frac{4\pi d}{\lambda}\cos(\theta)} \\ \vdots \\ e^{-j\frac{2\pi d(N-1)}{\lambda}\cos(\theta)} \end{bmatrix}. \tag{3.10}$$

The MVDR beamformer minimizes the total output power while maintaining distortionless response in the desired direction θ'. The optimization problem is

$$\min_{\mathbf{w}} \mathbf{w}^H \mathbf{R} \mathbf{w} \quad \text{subject to} \quad \mathbf{w}^H \mathbf{a}(\theta) = 1,$$

where $\mathbf{w}$ is the beamforming weight vector of size N and $\mathbf{R}$ is the covariance matrix of the received signal of size $N \times N$. The optimal MVDR beamformer weights are given by

$$\mathbf{w}_{\text{MVDR}} = \frac{\mathbf{R}^{-1}\mathbf{a}(\theta)}{\mathbf{a}^H(\theta)\mathbf{R}^{-1}\mathbf{a}(\theta)}.$$

The denominator $\mathbf{a}^H(\theta)\mathbf{R}^{-1}\mathbf{a}(\theta)$ ensures unit gain in the desired direction. Algorithm 3 presents the steps of beamforming to a target direction θ using MVDR. Repeat the algorithm for all desired spatial angles and identify the directions corresponding to local maxima in magnitude, which represent the potential source directions.

3.5 Channel Impulse Response Estimation

In Sect. 2.1, we have discussed aerial and structure bone acoustic channels and found that they have very different characteristics. For example, acoustic dispersion is more pronounced in solid materials, while the properties of aerial acoustic channels are greatly influenced by temperature, humidity, and air pressure. To estimate the channel impulse response (CIR) of either channel, a known reference signal (called *pilot signal*) is transmitted, and the resulting received signal is recorded. Various waveforms can be used as the pilot signal, including chirp, impulse signals, pseudo-random binary sequences such as maximum length sequences (MLS), or M-sequence as described in Sect. 2.3.

One straightforward approach to estimate CIR is by deconvolving the received signal with the transmitted pilot signal. Although it can be efficiently implemented in the frequency domain, this approach ignores noise spectrum and leads to large

Algorithm 3: MVDR beamforming for ULA with N elements

Require: Received signal matrix $\mathbf{x}(t)$, desired direction AoA θ, array geometry (spacing d, wavelength λ).

Ensure: Beamformer weights $\mathbf{w}_{\text{MVDR}}$ and output $y(t)$.

1: Compute the covariance matrix $\mathbf{R}$ using:

$$\mathbf{R} = \frac{1}{T}\sum_{t=1}^{T} \mathbf{x}(t)\mathbf{x}^H(t).$$

2: Calculate the steering vector $\mathbf{a}(\theta)$ for the given angle of arrival θ according to Eq. (3.10).

3: Calculate the MVDR weights using:

$$\mathbf{w}_{\text{MVDR}} = \frac{\mathbf{R}^{-1}\mathbf{a}(\theta)}{\mathbf{a}^H(\theta)\mathbf{R}^{-1}\mathbf{a}(\theta)}.$$

4: Filter the received signal $\mathbf{x}(t)$ with the MVDR weights:

$$y(t) = \mathbf{w}_{\text{MVDR}}^H \mathbf{x}(t).$$

5: **return** the beamformer weights $\mathbf{w}_{\text{MVDR}}$ and the output signal $y(t)$.

errors at frequency components in the pilot signal with very low magnitudes. Thus, it is only suitable for broadband pilot signals. Additionally, the frequency domain representation fails to shed much light on the geometric properties of the channel, such as the number of paths and their respective delays. As detailed in Chap. 6, these properties are useful in many acoustic sensing applications. To obtain the time domain CIR, one can perform an inverse discrete Fourier transform (IDFT) to the resulting spectral channel response.

Alternatively, under the assumption of LTI and FIR channels, the relation between pilot and received signals is given by Eq. (3.1). The values $h[0], h[1], \ldots, h[L-1]$ are called the taps of the FIR channel, and each tap corresponds to a specific delay and its associated weight (coefficient). To estimate the taps, we formulate the system in matrix form,

$$\mathbf{y} = \mathbf{X}\mathbf{h} + \mathbf{w},$$

where $\mathbf{y}$ is the received signal vector of length N ($N > L$), $\mathbf{h}$ is the channel tap vector, and $\mathbf{X}$ is a lower triangular matrix of the pilot sequence $x[n]$ given by

$$\mathbf{X} = \begin{bmatrix} x[0] & 0 & \cdots & 0 \\ x[1] & x[0] & \cdots & 0 \\ \vdots & \vdots & \ddots & \vdots \\ x[N-1] & x[N-2] & \cdots & x[N-L] \end{bmatrix}.$$

The MMSE (minimum mean square error) estimation method, also called *FIR Wiener filter*, finds optimal tap weights by using the statistics of the input and output signals by minimizing the mean square error between the estimated filtered signal and the output signals. The MMSE estimate of the channel impulse response $\hat{\mathbf{h}}$ is given by

$$\hat{\mathbf{h}}_{\mathrm{MMSE}} = \mathbf{R}_{xx}^{-1}\mathbf{v}.$$

Here, $\mathbf{R}_{xx} = \mathbb{E}[\mathbf{x}\mathbf{x}^H]$ is the covariance matrix of the transmitted signal:

$$\begin{bmatrix} R_x[0] & R_x[1] & \cdots & R_x[L-1] \\ R_x[1] & R_x[0] & \cdots & R_x[L-2] \\ \vdots & \vdots & \ddots & \vdots \\ R_x[L-1] & R_x[L-2] & \cdots & R_x[0] \end{bmatrix},$$

and $\mathbf{v} = \left[R_{xy}[0], R_{xy}[1], \ldots, R_{xy}[L-1]\right]^T$, where $R_{xy}[m] = E\{x[n]y[n+m]\}$.

Since in practice the covariances can only be approximated by finite samples, and computing matrix inversion is expensive, the least mean squares (LMS) algorithm is often implemented to approach the optimal Wiener filter through gradient descent iteratively as detailed in Algorithm 4.

Algorithm 4: Least mean squares (LMS) algorithm

Require: Observed input vector $\mathbf{x}(n)$, output signal $y(n)$, step size $\mu > 0$, and initial weight vector $\mathbf{h}[0]$.

Ensure: Updated weight vector $\mathbf{h}(n+1)$.

1: **Filter output:**

$$\hat{y}(n) \leftarrow \mathbf{h}^T(n)\mathbf{x}(n),$$

where $\mathbf{h}^T(n)$ is the transpose of the weight vector and $\mathbf{x}(n)$ is the input signal vector.

2: **Error signal:**

$$e(n) \leftarrow y(n) - \hat{y}(n).$$

3: **Weight update rule:**

$$\mathbf{h}(n+1) \leftarrow \mathbf{h}(n) + \mu e(n)\mathbf{x}(n),$$

where μ is the step size (a small positive constant).

3.6 Acoustic Simulation

Acoustic simulation leverages physics-based computational models to predict how sound propagates and interacts within a given environment. These tools are essential for designing and conducting early-stage testing of acoustic communication and sensing algorithms. They offer significant advantages due to their repeatability, controllability, and the ability to avoid the cost and time associated with building physical prototypes or conducting experiments. Recently, with the rise of data-intensive machine learning models, acoustic simulations have become a valuable tool for generating large-scale labeled datasets at a low cost. However, depending on the modeling simplifications applied, these simulations may compromise fidelity in favor of computational efficiency. In this section, we explore mainstream acoustic simulation methods, focusing on their applicability to active acoustic sensing.

3.6.1 Wave-Based Methods

Acoustic signals are waves that propagate in a medium. The 3D pressure field that acoustic waves generate satisfies the wave equation:

$$\nabla \cdot \left(\frac{1}{\rho(\mathbf{r})} \nabla p(\mathbf{r}, t) \right) - \frac{1}{c^2(\mathbf{r})} \frac{\partial^2 p(\mathbf{r}, t)}{\partial t^2} = -\delta(\mathbf{r} - \mathbf{r}_0) f(t), \tag{3.11}$$

where $c(\mathbf{r})$ is the spatially varying speed of sound, $\rho(\mathbf{r})$ is the density of the medium at point $\mathbf{r}$, $p(\mathbf{r}, t)$ is the pressure field, and $f(t)$ is the source time function. This equation governs wave propagation in environments with nonuniform media, such as complex structures like rooms or outdoor spaces with obstacles, and air with varying temperature. Green's function $G(\mathbf{r}, t; \mathbf{r}_0, t_0)$ is the solution to the acoustic wave equation for a point source that emits a pulse at time t_0 at position $\mathbf{r}_0$. Green's function can be expressed in the frequency domain by Fourier transforming the time domain wave equation. This leads to the *Helmholtz equation*:

$$\nabla \cdot \left(\frac{1}{\rho(\mathbf{r})} \nabla p(\mathbf{r}, \omega) \right) + \frac{\omega^2}{c^2(\mathbf{r})} p(\mathbf{r}, \omega) = -\delta(\mathbf{r} - \mathbf{r}_0). \tag{3.12}$$

Here, $p(\mathbf{r}, \omega)$ is the frequency domain pressure field and ω is the angular frequency. The frequency domain Green's function is useful for steady-state or harmonic source problems.

Green's function depends on factors such as boundary conditions (e.g., walls, reflecting or absorbing surfaces), medium's properties, and source-receiver geometry including the effects of diffraction, reflection, and refraction. With the exception of free space, Green's function cannot be derived analytically and has to be calculated numerically. Numerical methods, such as finite difference time domain

(FDTD), boundary element methods (BEM), or finite element methods (FEM), are often employed to model complex acoustic channels. BEM is often used in acoustic simulations in semi-infinite domains (e.g., around objects in an open or large space) and homogeneous medium. Compared to FEM, which discretizes the entire volume of a domain, BEM only discretizes the boundary of the domain and thus tends to have lower computation complexity. However, BEM assumes reasonably smooth surfaces. The presence of sharp corners and irregular boundaries can significantly impact its performance, leading to singularities in the field variables, inaccuracies, or divergence. The computational complexity of BEM is influenced by mesh density of the boundary; as density increases, the system matrix grows in size, leading to greater computational complexity. Commercial tools for acoustic simulations such as COMSOL implement FEM, BEM, and the hybrid of the two [2].

3.6.2 Geometrical Acoustics

Geometrical acoustics, also known as ray acoustics, approximates the propagation of acoustic waves as rays. For a good approximation, several conditions must hold: (1) The wavelength of the sound must be much smaller than the characteristic dimensions of the environment or obstacles, such as walls, openings, or objects, (2) the source and observer should be far enough from each other (relative to the wavelength), (3) diffraction is negligible, and 4) the medium is homogeneous. Therefore, it is suitable for high-frequency, large-scale problems, such as architectural acoustics and environmental noise studies [7]. Similar to geometrical optics, geometrical acoustics allows acoustic simulations through *ray tracing* and *image source* methods, which are generally of much lower computation complexity than FEM and BEM.

In ray tracing, sound propagation is simulated by tracing rays emitted from the sound source in different directions. These rays reflect off surfaces according to the laws of reflection, allowing multiple reflections and interactions with complex geometries, since each ray can reflect multiple times, and tracing multiple paths can be computationally intensive, especially in highly reflective environments. For a point source, ignoring atmospheric absorption, sound intensity decreases with the square of the distance in air, or equivalently, the sound pressure level (SPL) decreases by about 6 dB each time the distance from the source doubles. The absorption coefficient of indoor surfaces ranges from 0.01 (metal), 0.05–0.10 (concrete, brick, glass) to 0.80–0.90 for curtain and carpet, and 0.6–0.8 for humans at 20 KHz [3]. Therefore, it is often sufficient to limit the number of reflections to a small value (e.g., 3).

The image source method simulates reflections by creating “image" sources for each reflective surface. Each image source represents a virtual sound source positioned as if the sound had reflected off that surface. From one virtual sound source, additional (higher order) virtual sound sources can be further created. The sound intensity at a receiver is calculated from the sum of received signals from

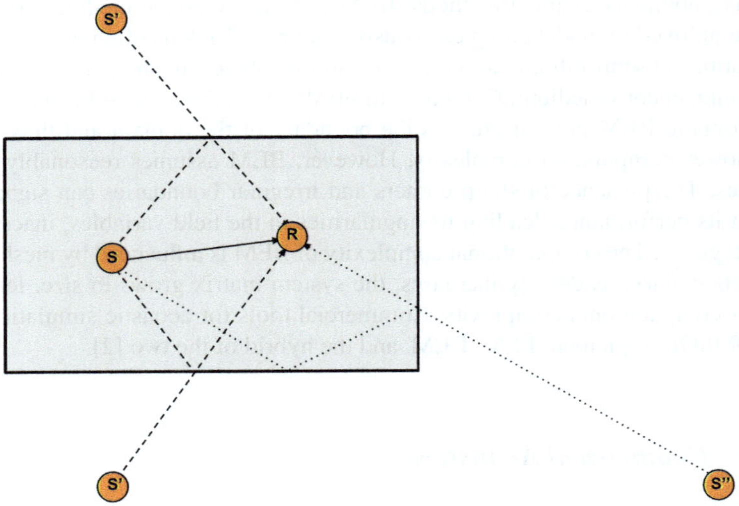

Fig. 3.9 Illustration of ray tracing and image source methods. S and R are source and receiver, respectively. S' and S'' are first- and second-order virtual image sources. The direct path, first-order reflected paths, and second-order reflected path are depicted in solid, dashed, and dotted lines, respectively

each virtual source following direct paths. From the discussion, we can see that if all reflective surfaces are accounted for, the two methods (ray tracing and image source) yield the same result. However, the latter is typically used to simulate scenes with regular geometries and relatively few reflective surfaces such as room acoustics. An illustration of ray tracing and image source methods is provided in Fig. 3.9.

Several open-source acoustic simulators are available that implement ray tracing or image source methods. As of the writing, popular ones include Pyroomacoustics [16] and RAVEN (Room Acoustics Visualizer for Engineers) [5].

3.7 Summary

In this chapter, we introduce the fundamental building blocks of acoustic sensing. Since acoustic signals are a type of signal, many techniques from signal processing and estimation theory are directly applicable. However, when developing active acoustic sensing applications on commodity devices, limitations and characteristics of these devices have to be taken into account. Implementing signal processing algorithms in practical systems faces nontrivial challenges such as synchronization, system delay, audibility, etc.

References

1. C. Cai et al., Boosting chirp signal based aerial acoustic communication under dynamic channel conditions. IEEE Trans. Mob. Comput. **21**, 1–11 (2021)
2. COMSOL Multiphysics, Introduction to COMSOL multiphysics®, in *COMSOL Multiphysics, Burlington, MA, accessed Feb 9, 2018* (1998), p. 32
3. T.J. Cox, P. D'Antonio, *Acoustic Absorbers and Diffusers: Theory, Design and Application*, 2nd edn. (Taylor & Francis, New York, 2009). ISBN: 978-04-15471-74-9
4. X. Cui, K. Yu, S. Lu, Evolutionary TDOA-based direction finding methods with 3-D acoustic array. IEEE Trans. Instrument. Measure. **64**(9), 2347–2359 (2015)
5. Institute of Technical Acoustics RWTH Aachen University. *RAVEN (Room Acoustics Visualizer for Engineers)*. https://www.virtualacoustics.org/RAVEN/ (2023)
6. K. Itoh, Analysis of the phase unwrapping algorithm. Appl. Opt. **21**(14), 2470–2470 (1982)
7. H. Kuttruff, *Room Acoustics* (CRC Press, Boca Raton, 2016)
8. P. Lazik, A. Rowe, Indoor pseudo-ranging of mobile devices using ultrasonic chirps, in *Proceedings of the 10th ACM Conference on Embedded Network Sensor Systems*. SenSys '12 (2012)
9. P. Lazik et al., ALPS: A bluetooth and ultrasound platform for mapping and localization, in *Proceedings of the 13th ACM Conference on Embedded Networked Sensor Systems*. SenSys '15 (2015)
10. K. Liu, X. Liu, X. Li, Guoguo: Enabling fine-grained indoor localization via smartphone, in *Proceeding of the 11th Annual International Conference on Mobile Systems, Applications, and Services*. MobiSys '13 (2013)
11. J. Liu et al., Snooping keystrokes with Mm-level audio ranging on a single phone, in *Proceedings of the 21st ACM MobiCom* (2015), pp. 142–154
12. W. Mao, J. He, L. Qiu, CAT: High-precision acoustic motion tracking, in *Proceedings of the 22nd Annual International Conference on Mobile Computing and Networking*. (2016), pp. 69–81
13. R. Nandakumar et al., FingerIO: Using active sonar for fine-grained finger tracking, in *Proceedings of the 34th ACM SIGCHI* (2016), pp. 1515–1525
14. C. Peng et al., Beepbeep: A high accuracy acoustic ranging system using cots mobile devices, in *Proceedings of the 5th ACM SenSys* (2007), pp. 1–14
15. K.M.M. Prabhu, *Window Functions and Their Applications in Signal Processing* (Taylor & Francis, Milton Park, 2014)
16. R. Scheibler, E. Bezzam, I. Dokmanic´, *Pyroomacoustics: A Python Package for Room Acoustics Simulations*. https://github.com/LCAV/pyroomacoustics. MIT License. Implements image source method for room acoustics simulations (2018)
17. M. Uddin, T. Nadeem, RF-Beep: A light ranging scheme for smart devices, in *Proceedings of IEEE PerCom* (2013), pp. 114–122
18. H.L. Van Trees, *Optimum Array Processing: Part IV of Detection, Estimation, and Modulation Theory* (Wiley, Hoboken, 2002)
19. Y.-T. Wang et al., ARABIS: An asynchronous acoustic indoor positioning system for mobile devices, in *2017 IEEE International Conference on Indoor Positioning and Indoor Navigation (IPIN 2017)* (2017)

Chapter 4
Aerial Acoustic Communication

4.1 Aerial Acoustic Communication

Airborne acoustic communication (AAC) allows nearby devices to exchange information wirelessly, offering several benefits over existing short-range communication technologies like near-field communication (NFC) and Bluetooth. Firstly, AAC is a lightweight solution with broad device compatibility; any IoT device equipped with a microphone or speaker can establish a communication link. Conversely, other wireless methods often necessitate specialized hardware, leading to additional expenses. Secondly, AAC exhibits low overhead, as it does not require the scanning or pairing procedures that are essential for Bluetooth devices. Thirdly, AAC offers enhanced security due to the limited propagation range of acoustic signals and their inability to easily penetrate walls, making eavesdropping challenging. These advantages have garnered the interest of researchers and technology companies, resulting in increased development and adoption of AAC.

The transmission of information via modulated radio signals has a long history in wireless communication systems. Similarly, underwater acoustic communication is a mature field with established real-world applications. The principles and techniques from wireless and underwater acoustic communication can potentially be adapted for use in AAC. However, the distinct nature of acoustic signals and the characteristics of aerial acoustic channels necessitate specialized design considerations for AAC.

In wireless communication, for example, frames commonly include known preamble sequences to aid in channel estimation. This allows for compensation of distortions in the carrier signals, ultimately reducing the bit error rate (BER). However, this compensation is only effective if performed within a coherent time interval, during which the channel response can be considered relatively constant. Achieving similar compensation in AAC presents two main challenges. Firstly, the coherent channel time is short, typically limited to a few milliseconds in a standard

R. Zheng, C. Cai, *Acoustic Sensing on Commodity Devices and its Applications*, Wireless Networks, https://doi.org/10.1007/978-3-031-96875-4_4

indoor setting (e.g., 8 ms as reported in [7]), making use of long training sequences impractical due to the low data rate. Secondly, channel compensation often requires complex computations to be performed in real time, which can be challenging for IoT devices with limited resources.

4.1.1 A Measurement Study

To gain a firsthand understanding of acoustic signal stability, we conducted a series of measurements to evaluate crucial parameters, specifically frequency, phase, and amplitude, which are potentially useful for modulation. The findings from these measurements will guide the selection of suitable modulation techniques and the design of resilient packet structures.

The experimental setup involved a desktop computer continuously playing a 20 kHz pure tone, which was output through a 3.5 mm audio jack and amplified by a 20 W class-A power amplifier to reduce distortion. This amplified signal was then broadcast using a 10-inch speaker. A Samsung Galaxy S5 smartphone, located 20 cm from the speaker, recorded the transmitted signal using a custom application. This application, built upon the Android AudioTrack class, captured unprocessed PCM audio data, bypassing any internal software filtering. The recording lasted for 40 seconds and was performed in an office setting. The average background noise level was around 44 dB, with intermittent peaks surpassing 76 dB due to environmental noise sources such as human speech, keyboard strokes, and desk knocks. The results of these measurements are shown in Fig. 4.1.

Figure 4.1a illustrates the time-frequency spectrum of the initial signals, from which the prominent 20 kHz pure tone signal is clearly discernible. Given that the majority of noise resides below 5 kHz, we employed a 30th-order high-pass FIR filter with a cutoff frequency of 5 kHz to mitigate it. The use of an FIR filter was chosen to preserve phase information. As depicted in Fig. 4.1b, the filtering process effectively removed low-frequency noise; however, some broadband noise persists between 20∼30 s. Despite this, the data presented in Fig. 4.1b indicates that the acoustic signal's frequency remains relatively stable, suggesting its potential for modulation techniques.

Figure 4.1c and d depict the amplitude and phase characteristics of the filtered signals. The amplitude envelope, extracted using the Hilbert transform and shown in Fig. 4.1c, exhibits considerable variation, ranging from a peak of 1 to a near-zero minimum. This substantial fluctuation renders amplitude modulation an unfavorable choice for AAC due to the instability it introduces in signal encoding. To analyze the phase, as presented in Fig. 4.1d, we first divided the filtered signal into segments of 48,000 samples each. Subsequently, we applied the FFT to each segment and extracted the phase of the 20 kHz component, which corresponds to the 20,001st index in the FFT spectrum. The results in Fig. 4.1d reveal a nearly linear increase in phase over time, with phase shifts of up to 10 radians observed over a 10-second interval. This inherent phase drift explains the infrequent use of phase-based

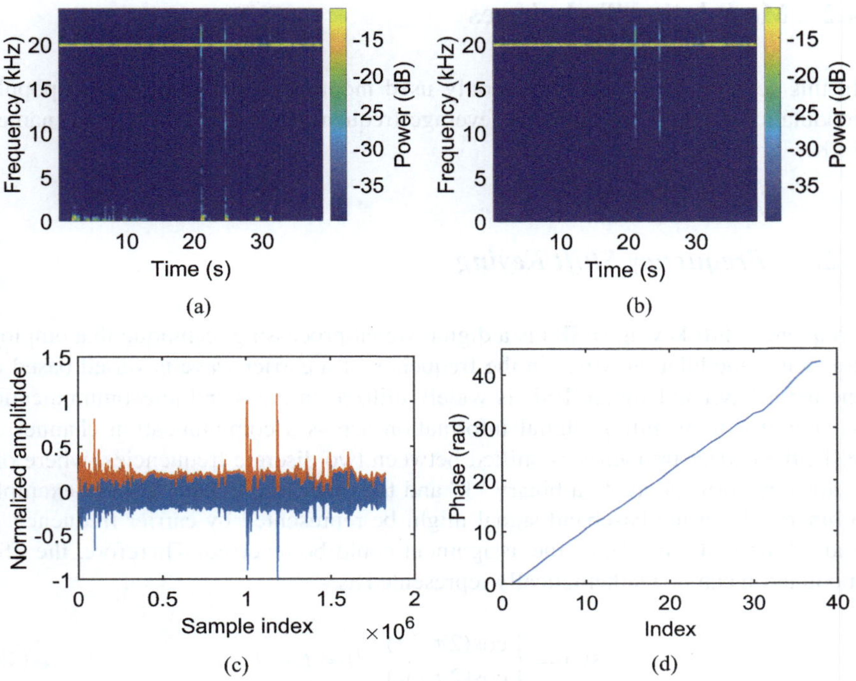

Fig. 4.1 Measurement results. (**a**) Time frequency of original signals. (**b**) Time frequency of filtered signals. (**c**) Amplitude of filtered signals. (**d**) Phase of filtered signals

modulation techniques in AAC, as they necessitate precise phase synchronization, a challenging requirement in real-world acoustic environments.

The instability of both amplitude and phase in acoustic communication stems from various factors. Regarding amplitude, a major source of unreliability is background interference. Furthermore, the automatic gain control (AGC) present in acoustic front-end systems adds to these challenges. When the AGC is operating, abrupt loud noises—like a sharp knock on a desk—can initiate gain adjustments to avoid audio clipping. Consequently, the AGC decreases its gain, which in turn reduces the recorded intensity of other signals. This interplay between AGC and background interference makes modulation based on amplitude unsuitable for AAC. The main reason for significant phase deviations is the presence of carrier frequency offset (CFO) or sampling frequency offset (SFO). These offsets occur because the sampling rates of different devices are not perfectly aligned, resulting in accumulated errors over time. In radio communication systems, CFO and SFO are usually corrected using channel equalization techniques. However, implementing such compensation in acoustic systems is not feasible due to limitations in computational resources and hardware capabilities. Consequently, AAC systems commonly employ frequency modulation techniques, which are inherently more robust against these types of distortions.

4.2 Modulation Techniques

In this section, we present commonly used modulation techniques. One popular modulation method is to directly leverage frequency to carry information, namely FSK.

4.2.1 Frequency Shift Keying

Frequency shift keying (FSK) is a digital signal processing technique that employs frequency modulation, wherein the frequency of a carrier wave is varied based on the input baseband signal. FSK is widely utilized in radio and telecommunication systems for transmitting digital information across a communication channel. In FSK, the carrier frequency is shifted between two discrete frequencies, where one frequency corresponds to a binary "1" and the other to a binary "0." For example, a binary "1" in the baseband signal might be represented by carrier frequency f_1 and a binary "0" by f_2, or the assignment could be reversed. Therefore, the FSK modulation can be mathematically represented as

$$\mathrm{s(t)} = \begin{cases} \cos(2\pi f_1 t) \\ \cos(2\pi f_2 t) \end{cases}, 0 < t < T, \tag{4.1}$$

where $s(t)$ denotes a carrier whose duration is T, and f_1 and f_2 are the two carrier frequencies. Figure 4.2 shows an example of FSK modulated signal. In the figure, the baseband signal is [1, 1, 1, 0, 0, 1, 1] and $f_1 = 19\,\text{kHz}$ and $f_2 = 22\,\text{kHz}$. The duration of a carrier lasts for 40ms. By performing a spectral analysis, say FFT, the embedded messages can thus be decoded.

As shown in Fig. 4.2, frequency leakage is evident at the interface of consecutive intervals when the input symbols change. The sharp amplitude transition between the two carriers is the cause of this issue. One potential solution to reduce this frequency leakage is to employ window functions, for example, a Hanning window,

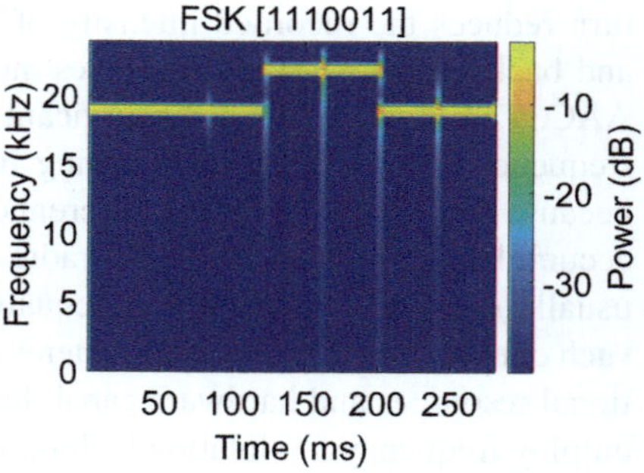

Fig. 4.2 Time-frequency analysis of a FSK signal

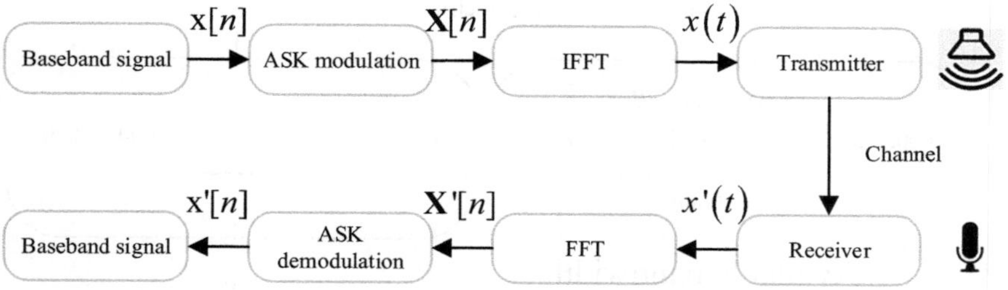

Fig. 4.3 Procedures for OFDM modulation and demodulation

on the carrier. FSK is advantageous due to its straightforward implementation. However, a drawback of FSK is its relatively low achievable data rate. Consequently, this low data rate necessitates the use of OFDM.

4.2.2 OFDM

OFDM Basics

Orthogonal frequency division multiplexing (OFDM) is a sophisticated modulation technique frequently employed in fourth-generation (4G) and fifth-generation (5G) mobile communication networks (Fig. 4.3). As previously discussed in Chap. 2, the direct application of conventional OFDM modulation to acoustic environments is challenging due to factors such as the limited coherence time and significant phase noise. To address these limitations and enhance the robustness of acoustic communication, OFDM is often integrated with the amplitude shift keying (ASK) for data modulation.

The procedures for OFDM encoding and decoding mechanism are as follows. Given a baseband signal of length N $\mathbf{x}[n] \in \mathcal{R}^N$, we first perform ASK modulation on $\mathbf{x}[n]$. The rational for the adoption of ASK has been elaborated in Fig. 4.1d as phase is time variant even under static channels and hence is unreliable. Before the ASK modulation, we can also apply channel coding techniques [12] on baseband signals to add redundancy so as to combat noises. After this ASK modulation, the baseband signal $\mathbf{x}[n]$ is treated as a frequency domain signal $\mathbf{X}[n]$. The inverse fast Fourier transform (IFFT) is applied to convert $\mathbf{X}[n]$ into ready-to-be-transmitted time domain signals $x(t)$. To ensure that $x(t)$ is a real signal, $\mathbf{X}[n]$ must be Hermitian symmetric, namely $X[-n] = X^*[n]$, before performing the IFFT, where $X^*[n]$ denotes the complex conjugate of $X[n]$. The detailed procedures to transform ASK modulated signals into OFDM signals are illustrated in Fig. 4.4.

The process of transforming ASK modulated signals into OFDM signals converts a single sideband signal (SSB) into a double sideband (DSB) one with symmetrical spectral. Let the bandwidth of $\mathbf{X}[n]$ be B_N and the available bandwidth be B_W. The number of points in IFFT is given by $\frac{NB_W}{B_N}$. At the sampling frequency is F_s,

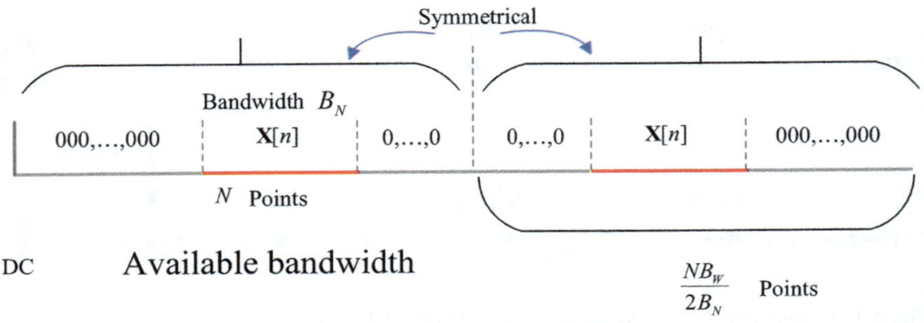

Fig. 4.4 From ASK to OFDM

and the frequency resolution or the carrier spacing is thus $\frac{F_s B_N}{2N B_W}$. To construct a DSB frequency domain signal, the N-sample sequence $\mathbf{X}[n]$ must be symmetrically mapped to a sequence of length $\frac{N B_W}{B_N}$. The starting index I of the lower half sequence determines the occupied bandwidth $[I \times \frac{F_s B_N}{2N B_W}, (I + N - 1) \times \frac{F_s B_N}{2N B_W}]$ of the ASK signals.

An Illustrative Example

Let us consider the architecture of an inaudible communication system designed for an acoustic device operating with a sampling frequency of 48 kHz. The modulation process can utilize a bandwidth spanning from 18 to 22 kHz. To achieve robust OFDM transmission, a carrier spacing of 100 Hz is chosen, which yields a frequency resolution of 100 Hz. With this setup, each OFDM symbol can encode $4000/100 = 40$ bits, implying that 40 subcarriers are employed for data modulation. Consequently, we define $N = 40$. The total number of points for the OFDM signal in the frequency domain is calculated as $2 * 40 * 24/4 = 480$.

Subsequently, a 40-bit sequence is mapped onto a sequence of 240 samples. Given that the effective frequency range spans from 18 to 22 kHz, the corresponding lowest and highest frequency indices are 180 (calculated as 18000/100) and 219 (calculated as 22000/100 - 1), respectively. To achieve a double sideband (DSB) signal, the ASK modulated symbols are mirrored and positioned in the upper half of the frequency sequence, specifically at indices from 261 to 300. For this illustration, we consider an ASK modulation scheme where a binary "1" is mapped to an amplitude of 1, and a binary "0" is mapped to an amplitude of 0. The process of transforming ASK signals into the OFDM frequency domain representation is depicted in Fig. 4.5a through b, while Fig. 4.5c illustrates the time-frequency domain representation.

As the final operation, the frequency domain signals are converted into a time domain waveform by employing the inverse fast Fourier transform (IFFT). This process yields an inherently real-valued signal, a consequence of the spectrum's Hermitian symmetry, which permits direct transmission.

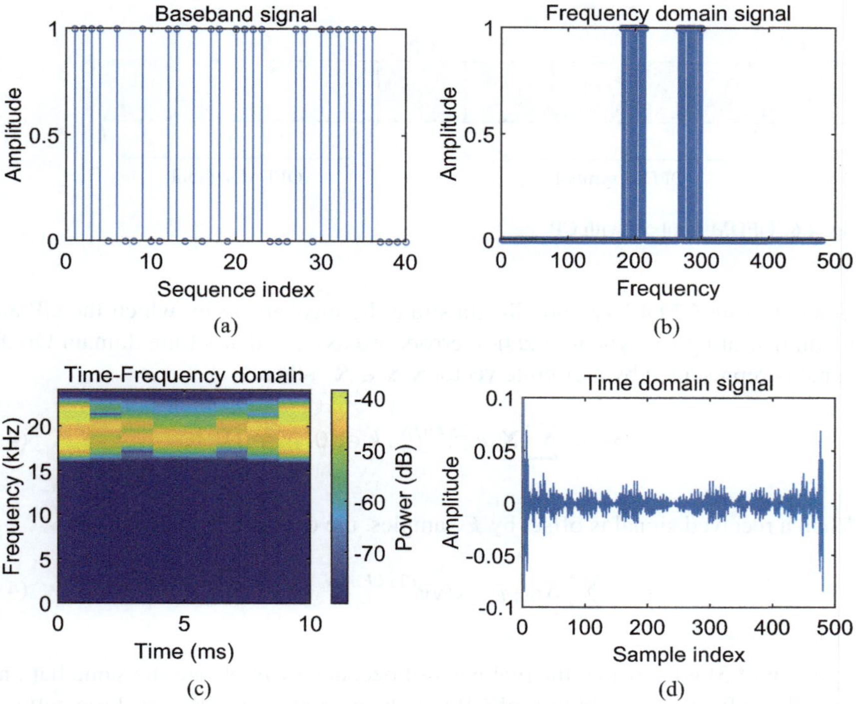

Fig. 4.5 OFDM modulation: signal representations. (**a**) Baseband signals. (**b**) Frequency domain OFDM signals. (**c**) Time-frequency domain OFDM signals. (**d**) Time domain OFDM signals

Practical Considerations

Despite its capacity for high data rates, orthogonal frequency division multiplexing (OFDM) is susceptible to the peak-to-average-power ratio (PAPR) issue [11]. The PAPR problem refers to a phenomenon where a limited number of samples exhibit significantly high magnitudes. This characteristic is evident in Fig. 4.5d, which illustrates that the magnitude of the initial and final few samples is considerably greater than that of the remaining samples. This PAPR issue can result in frequency leakage or audible noise, even though the modulation is performed within the inaudible frequency spectrum. While radio-based wireless communication systems offer solutions to alleviate this problem, these methods are generally not suitable for acoustic-enabled IoT devices because of their considerable complexity. Presently, effective solutions for the PAPR problem within acoustic domains remain largely unexplored.

Furthermore, it is crucial to acknowledge that, in practical implementations, an OFDM symbol frequently incorporates a cyclic prefix (CP) to reduce the impact of synchronization errors. This aspect is especially significant in the design of acoustic signals, where achieving precise synchronization presents inherent difficulties [8]. As depicted in Fig. 4.6, a cyclic prefix involves repeating the final

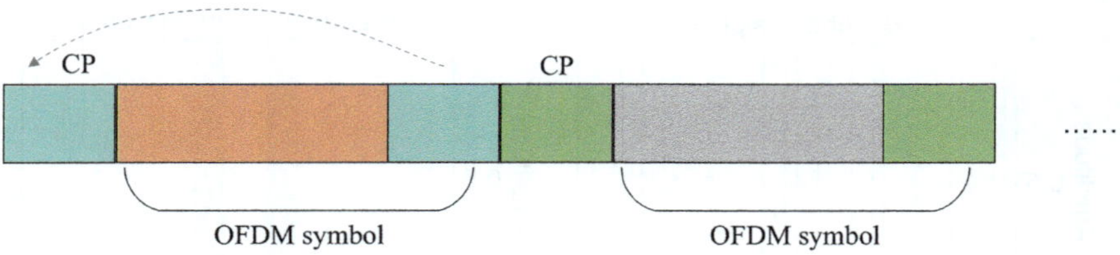

Fig. 4.6 OFDM symbols with CP

segment of an OFDM symbol. To illustrate the mechanism by which the CP aids in compensating for synchronization errors, we assume that a time domain OFDM signal is represented by a discrete vector $\mathbf{x}$, $\mathbf{x} \in \mathcal{R}^N$, then:

$$\mathbf{x}_k = \sum_i \mathbf{X}_n e^{i2\pi kn/N}, k \in [0, N-1]. \tag{4.2}$$

When a received signal is offset by E samples, the decoded $\mathbf{x}_k$ becomes [8]

$$\mathbf{x}_{k+E} = \sum_i \mathbf{X}_{(n+E)modN} e^{i2\pi kEn/N}, k \in [0, N-1]. \tag{4.3}$$

Equation (4.3) tells us that the real parts of decoded symbols are the same but only phase has offsets. The addition of CP may be crucial as perfect synchronization is hard to achieve in acoustic domain.

In conclusion, although OFDM provides excellent spectral efficiency, its encoded symbols are quite vulnerable to propagation loss, which restricts its effective communication range to near-field applications spanning only a few centimeters. To achieve communication ranges extending to the meter scale, chirp spread spectrum (CSS) presents a promising alternative, capitalizing on its inherent resilience to attenuation and multipath interference.

4.2.3 *Chirp Spread Spectrum*

CSS is suitable for communication at long distances. As mentioned in Chap. 2, the chirp signal can be expressed as

$$s(t) = A\sin(2\pi f_{\min}t + \pi kt^2 + \phi), \tag{4.4}$$

where A denotes the magnitude of chirp signal, $f_{\min}$ is the initial frequency, k is the modulation coefficient, and ϕ denotes the initial phase. Unlike long range (LoRa), a wireless communication technology designed for transmitting small amounts of data over extensive distances, the modulation of messages in the acoustic domain rarely employs $f_{\min}$. Instead, the parameter k is more commonly utilized for modulation due to the limited bandwidth available for acoustic communication.

The approach to leveraging chirp signals for message modulation in the acoustic domain differs fundamentally from that of OFDM. In OFDM, although inter-symbol interference exists, the interference from adjacent symbols is null at the sampling points. Consequently, each symbol in OFDM is considered orthogonal, as expressed by

$$s_i \bigotimes s_j \doteq 0, i \neq j, \tag{4.5}$$

where s_i and s_j denote distinct symbols, and $\bigotimes$ represents any possible operations between two symbols, for instance cross-correlation, equal to zero.

However, establishing orthogonality among signals modulated using chirp spread spectrum (CSS) is a complex task. To overcome this difficulty, the researchers in [9] introduced orthogonal chirp division multiplexing (OCDM), a modulation technique founded on the Fresnel transform that allows for the creation of orthogonal chirps fulfilling the conditions of Eq. (4.5). Despite this advantage, OCDM exhibits a high degree of sensitivity to synchronization offsets, which restricts its practical use. Consequently, the majority of systems utilize time division, where individual symbols are allocated to different time intervals. Nevertheless, this method introduces a fundamental trade-off: Achieving a higher data rate requires shortening the duration of each symbol. This reduction, in turn, lowers the processing gain associated with chirp-based carriers, eventually resulting in a higher bit error rate (BER).

Another approach to reduce the mutual interference between different symbols is by varying k. In [2, 3], the authors propose the concept of loose orthogonality. Let f and g be two different chirp carriers. They can be regarded as loose orthogonal if the following criterion is met:

$$\alpha = \frac{\max R_{fg}(\tau)}{\min\left(\max R_{ff}(\tau), \max R_{gg}(\tau)\right)} < \xi, \tag{4.6}$$

where ξ is a threshold value indicating the permissible degree of looseness. $R_{fg}(\tau)$ denotes the correlation between f and g given by

$$R_{fg}(\tau) = \int_T f(t)\, g(t-\tau)\, dt, \tag{4.7}$$

where T represents signal duration. In Eq. (4.6), α characterizes the degree of interference between f and g if transmitting at the same time. A smaller α is preferable as it is easier to isolate f or g from any mixture using a matched filter.

An initial approach involved using the sign of the modulation coefficient k to encode information, a method referred to as binary orthogonal keying (BOK). In this technique, a positive modulation coefficient k_0 encodes the binary digit "1" (an up-chirp), whereas its negative counterpart, $-k_0$, encodes the binary digit "0" (a down-chirp). Figure 4.7 provides an illustration of BOK modulation.

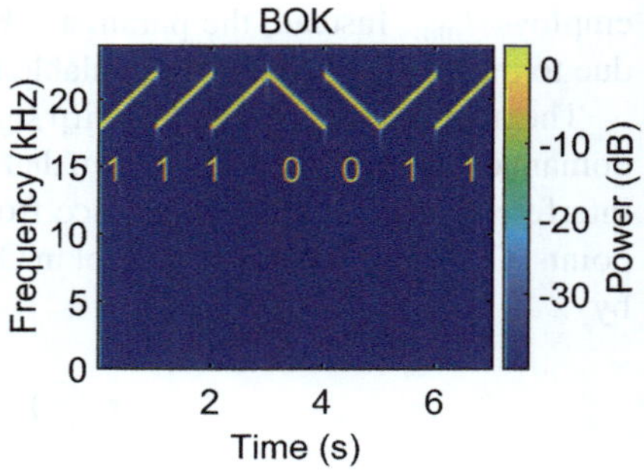

Fig. 4.7 BOK modulation example: using the sign of the modulation coefficient of chirp carrier to embed messages

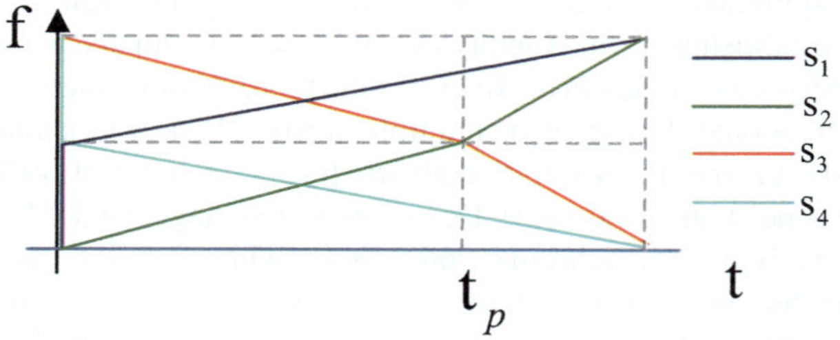

Fig. 4.8 QOK symbols: The introduction of "turning points" at different positions in chirps results in multiple orthogonal symbols [6]

Following this, research by Ka et al. [6] introduced the concept of "turning points" within the modulation coefficient k to modulate data, effectively enhancing the modulation order to quadrature orthogonal keying (QOK). As depicted in Fig. 4.8, symbols s_1 and s_2 both start as up-chirps and have the same bandwidth and duration. However, they are differentiated by their turning points occurring at $t = 0$ and $t = t_p$, respectively, leading to a form of loose orthogonality. The same principle is applicable to symbols s_3 and s_4.

Expanding on this idea, the authors of [2] further increased the modulation order, bypassing the limitations of modulation techniques based on time division. Their work revealed that, in addition to the sign of k, the frequency range could also be employed as an extra parameter for modulation. Although simultaneously using multiple carriers within the same time interval usually leads to inter-carrier interference (ICI), they showed that if this interference is kept sufficiently low and does not prevent the accurate detection of individual carriers, this method can be used to increase the modulation order. Based on this finding, they proposed a loosely orthogonal modulation technique that utilizes multiple carriers with different central frequencies, which can effectively improve data transmission rates from tens of bits per second (bps) to as high as 1 kbps (Fig. 4.9).

The demodulation of signals modulated with chirp spread spectrum (CSS) frequently requires considerable computational power, mainly because it relies on matched filtering. The function of a matched filter is to correlate the incoming

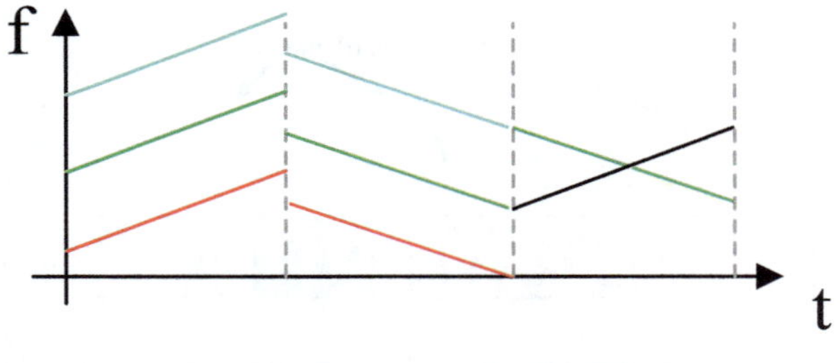

Fig. 4.9 The loose orthogonal modulation scheme: It overlaps multiple carriers within a single time duration to increase modulation order

signal $r(t)$ with a predetermined reference template $s_i(t)$. Detection of $s_i(t)$ occurs if the maximum magnitude of the correlation output, given by $\max(|R(\tau)|) = \max|\int_T r(t)s(t-\tau)dt|$, surpasses a specific threshold or if it is greater than the correlation results obtained with any other reference symbols $s_j(t)$, $j \neq i$, $s_i(t)$ is detected. In the case of high-order modulation schemes, each information symbol has its own unique reference template, requiring repetitive correlation calculations for every possible symbol.

4.3 From Modulation to Acoustic Frames

In this section, we outline the architecture of a suitable acoustic frame. Typically, such a frame is structured with three essential parts: a preamble, data symbols, and a guard interval positioned between sequential frames, as visualized in Fig. 4.10. The receiver uses the preamble to recognize the arrival of an acoustic frame. To achieve the highest possible detection sensitivity and improve the likelihood of detection, a chirp carrier is often utilized, occupying the entire permitted bandwidth and time duration. Modulation of the data symbols is performed using the techniques detailed in Sect. 4.2. This silent guard interval has two primary objectives: to mitigate the adverse effects of multipath propagation and to minimize inter-symbol interference (ISI).

In scenarios where both the transmitter and receiver are static and the communication distance is short, straightforward methods like matched filtering can effectively retrieve information symbols. However, when channel conditions are dynamic, further signal processing stages become necessary to maintain reliable performance. As previously mentioned, the limited coherence time inherent in acoustic channels renders commonly used techniques like channel equalization ineffective for acoustic communication (AAC). Several initial attempts have been made to address this problem. The researchers in [2] suggested a computationally efficient channel estimation method. The fundamental concept involves transmitting predetermined symbols and evaluating channel conditions based on the observed

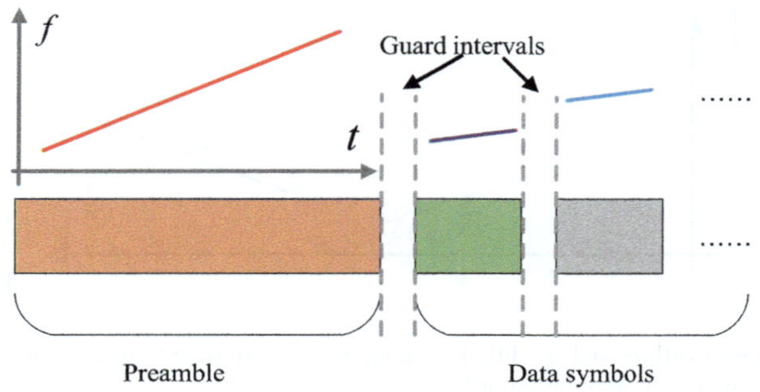

Fig. 4.10 A typical acoustic frame

decoded error rate. This approach is effective for handling moderate channel variations. In the work by Wang et al. [13], a pure tone signal is transmitted concurrently with OFDM symbols to estimate Doppler shifts caused by movement. These estimated shifts are then compensated for in the received signal to enhance the bit error rate (BER) of the decoded symbols.

Our prior discussion centered on aspects of the physical layer. Nevertheless, when multiple transmissions occur simultaneously, medium access control (MAC) becomes crucial for preventing a high number of collisions within acoustic channels. Time division multiple access (TDMA) is a commonly employed MAC protocol in acoustic systems owing to its straightforward nature and effectiveness. As an example, the work in [14] utilizes TDMA to manage the transmission timing of several acoustic anchors within an infrastructure-based acoustic localization system. The clocks of these anchors are synchronized using the network time protocol (NTP) through wireless communication links, enabling anchors within hearing distance to transmit in an orderly sequence. Although more complex MAC protocols could be adopted for AAC, they frequently lead to significant processing overhead or undesirable delays. In numerous situations, the physical separation of devices (such as by walls) inherently limits interference in acoustic channels, thus diminishing the need for intricate MAC protocols.

4.4 Open Problems and Research Challenges

Significant advancements have been made in the design of practical AAC systems. However, several open problems and research challenges persist, making it a continuously evolving field of study and development.

Limited Data Rate Aerial acoustic communication (AAC) typically offers data rates below 1 kbps, which considerably restricts its use in numerous real-world applications. For example, in-vehicle networking necessitates a minimum data rate of 10 kbps, rendering AAC inappropriate for such contexts. Possible strategies for enhancing the data rate include: (1) employing high-efficiency modulation techniques that are resilient to channel variations and background noise, and (2) utilizing spatial diversity by creating simultaneous communication links and using RAKE receivers to combine them [10]. Nevertheless, both of these methods require substantial computational resources, which are presently implemented using software.

Furthermore, the data rate attainable through acoustic means diminishes considerably with increasing distance. Once the separation between transceivers goes beyond a few meters, the effective data rate frequently falls to just tens of bps, assuming the communication link remains functional at these distances. Addressing the challenge of long-range, high-speed acoustic communication might be feasible with the use of acoustic metasurfaces [1]. These structures have the potential to act as RAKE receivers, effectively gathering incident acoustic waves and focusing them toward the receiver to improve signal quality [15]. In addition, acoustic metasurfaces present a possible approach for non-line-of-sight (NLoS) scenarios, enhancing signal reception in difficult environments.

Dynamic Channel Conditions The inherent variability of acoustic channels, encompassing Doppler effects and fluctuations in the distance between transceivers, presents a substantial obstacle to ensuring consistent link performance. These factors make the design of modulation schemes more complex, especially considering that real-time channel estimation is frequently not feasible. Moreover, multipath propagation and frequency-selective fading, which can be intensified by the characteristics of acoustic sensors, can significantly reduce achievable data rates. A potential solution lies in the use of orthogonal time-frequency space (OTFS) modulation [4, 5], a technique inherently robust to variations in time and frequency. However, the high computational demands of OTFS decoding currently limit its widespread use in AAC.

To lessen the effects of multipath propagation, microphone arrays, which are often included in smart speakers, can be utilized to improve signal reception. Another feasible strategy involves expanding the operating bandwidth into the ultrasound range, enabling AAC systems to leverage this extra bandwidth to effectively counteract channel dynamics.

References

1. B. Assouar et al., Acoustic metasurfaces. Nat. Rev. Mat. **3**(12), 460–472 (2018)
2. C. Cai et al., Boosting chirp signal based aerial acoustic communication under dynamic channel conditions. IEEE Trans, Mob. Comput. **21**(9), 1–11 (2021)
3. C. Cai et al., Reliable high throughput aerial acoustic communication for mobile network. IEEE Trans. Vehic. Technol. **73**(4), 5704–5716 (2024)
4. R. Hadani, A. Monk, OTFS: A new generation of modulation addressing the challenges of 5G (2018). arXiv preprint arXiv:1802.02623

5. R. Hadani et al., Orthogonal time frequency space modulation, in *2017 IEEE Wireless Communications and Networking Conference (WCNC)* (IEEE, Piscataway, 2017), pp. 1–6
6. S. Ka et al., Near-ultrasound communication for TV's 2Nd screen services, in *Proceedings of the 22Nd Annual International Conference on Mobile Computing and Networking*. MobiCom '16 (2016)
7. S. Ke et al., VSkin: Sensing touch gestures on surfaces of mobile devices using acoustic signals, in *Proceedings of the 24th ACM MobiCom* (2018), pp. 591–605
8. R. Nandakumar et al., FingerIO: Using active sonar for fine-grained finger tracking, in *Proceedings of the 34th ACM SIGCHI* (2016), pp. 1515–1525
9. X. Ouyang, J. Zhao., Orthogonal chirp division multiplexing. IEEE Trans. Commun. **64**(9), 3946–3957 (2016)
10. J.G. Proakis, *Digital Communications*, 5th edn. (McGraw-Hill, New York, 2007). ISBN: 978-00-72957-16-7
11. Y. Rahmatallah, S. Mohan, Peak-to-average power ratio re-duction in OFDM systems: a survey and taxonomy. IEEE Commun. Surv. Tutor. **15**(4), 1567–1592 (2013)
12. K.D. Rao, *Channel Coding Techniques for Wireless Communications*, 2nd edn. (Springer Publishing Company, Berlin, 2019). ISBN: 978-98-11505-60-7
13. Q. Wang et al., Messages behind the sound: Real-time hidden acoustic signal capture with smartphones, in *Proceedings of the 22nd Annual International Conference on Mobile Computing and Networking*. MobiCom '16 (2016)
14. Y.-T. Wang et al., ARABIS: An asynchronous acoustic indoor positioning system for mobile devices, in *2017 IEEE International Conference on Indoor Positioning and Indoor Navigation (IPIN 2017)* (2017)
15. H. Zhang et al., Optimizing multi-user indoor sound communications with acoustic reconfigurable metasurfaces. Nat. Commun. **15**(1), 1–10 (2024)

Chapter 5
Location Services

5.1 Ranging

Ranging (i.e., determining the distance between two devices) provides useful contextual information and finds applications in distance and size measurements, network management [27], proximity detection, and content sharing [5, 7]. Leveraging acoustic signals for ranging is an economical and convenient alternative to traditional measurement tools. Ranges are computed by multiplying the time of flight estimated using techniques in Sect. 3.3, with the propagation speed of acoustic signals, generally assumed to be constant and known a priori. *Ranging errors* are defined as the root mean square (RMS) of the differences between estimated results and the ground truth, while *ranging accuracy* is inversely proportional to the RMS error. Ranging performance heavily relies on the accuracy of the underlying timing estimation techniques such as robust onset detection, one-way or two-way ToA estimation, etc. Existing acoustic ranging solutions can be broadly categorized into: (1) *device-based*, which involves ranging between a pair of acoustic devices, and (2) *device-free*, namely, ranging from an acoustic device to a non-acoustic-emitting target. The latter when operated in air is also called **acoustic radar**. A comparison of different ranging approaches can be found in Table 5.1.

5.1.1 Device-Based Pairwise Ranging

BeepBeep [31] is a pioneer work that uses acoustic signals for precise ranging on commodity mobile devices based on ToA via a two-way sensing method (Sect. 3.3.2). It cleverly circumvents unknown system delays and clock synchronization by directly retrieving timestamps from acoustic samples themselves. To mitigate multipath effects, BeepBeep locates the earliest "sharp" peak after cross-correlating the received signals with reference signals. It reports centimeter-level

R. Zheng, C. Cai, *Acoustic Sensing on Commodity Devices and its Applications*,
Wireless Networks, https://doi.org/10.1007/978-3-031-96875-4_5

Table 5.1 Comparison of different ranging techniques

Method	Techniques	Waveform	BW (kHz)	Accuracy
BeepBeep [31]	Cross-correlation	Chirp signal	2–6	Avg. 1–2 cm within 10 m
RF-Beep [35]	One-way sensing	Tone	18	30 cm median error within 16 m
SwordFight [47]	Autocorrelation	Tone	0–11.025 or 0–16	Median error 2 cm within 2 m
DeepRange [23]	Deep learning	Chirp signal	18–22	A median of 1 cm within 4 m
BatMapper [49]	Cross-correlation with device constraints	Chirp signal	8–16 and 8–10	2 cm within 4 m
SAMS [32]	Phase-based timing	Chirp signal	11–21	Median 30 cm and a 90-quantile of 1 m within 5 m

(2–6 cm) ranging errors. However, the performance of BeepBeep can be degraded by the presence of strong NLOS paths. SwordFight [47] is another ranging solution that is designed for phone-to-phone mobile motion games. It improves upon BeepBeep in responsiveness, accuracy, and robustness. It utilizes the same one-way sensing mechanism as BeepBeep but differs in the transmitted waveforms and onset detection strategies. Specifically, it sends reference signals consisting of two copies of the same short pseudo-random sequence. Onset detection is done through autocorrelation where a receiver upon detecting the presence of the reference signal correlates the detected signal with a delayed version of the signal. To achieve more accurate onset detection, additional post-processing steps are applied including smoothing the autocorrelation peaks and small-scale cross-correlation by choosing a suitable search window size. SwordFight reports a median ranging accuracy of 2 cm with 12 Hz fresh rate in noisy environments.

Uddin and Nadeem in [35] mitigate the uncertain and variable system delays in acoustic playback and recording on commodity devices by a kernel space implementation and develop a stand-alone application called RF-Beep. In RF-Beep, ranging is done via ToA estimation following the one-way sensing approach illustrated in Fig. 3.3a, where RF signals are transmitted at the same time as an acoustic reference signal, and the arrival time of the RF signals is treated as the transmission time of the acoustic signal due to the negligible propagation time of the former in short distances. RF-Beep reports decimeter-level (up to 50 cm) ranging errors when two devices are within 16 m. Despite the impressive performance, the reliance on kernel modification creates a barrier to its wide adoption.

5.1.2 Acoustic Radar

Similar to RF radars, acoustic radars work by radiating acoustic energy in the environment and analyzing reflected echoes. In addition to ranging, it can be used in obstacle avoidance, mapping and imaging applications. Since the intensity of acoustic echoes (Watt per unit area) decreases sharply with the increase of distance d (e.g., in free space, proportional to $\frac{1}{d^2}$ [46]), the operational range of acoustic radars is often limited to meters or tens of meters. Multipath effects also pose additional challenges as it is often difficult to distinguish desired echoes bouncing off targeted objects from those experiencing multiple reflections or reflected off undesired objects. The evaluation metrics for ranging errors form acoustic radars are the same with the ones used in device-based approaches.

In [10], Graham et al. proposed a sonar sensor for depth sensing on smart phones utilizing the phone's microphone and rear speaker. A FMCW signal shaped by a Hanning window is transmitted. Ranging is estimated via cross-correlating received signals with the reference FMCW signal in frequency domain for fast computation and peak finding. The reported ranging errors are up to 12 cm within 40 m distances. Further improvement to ranging accuracy of acoustic radars is proposed in [23] using deep learning techniques. In this chapter, synthesized data is used to train a neural network to handle platform diversity, multipath effects, and background interference simultaneously. The resulting range error can be as low as 1 cm at a distance up to 4 m and is agnostic to various background noises, heterogeneous devices, etc.

Building upon ranging information, BatMapper [49] is among the first work that demonstrates the feasibility of commodity mobile devices to act as acoustic radars for indoor floor map construction. It exploits speaker-microphone distance constraints to construct a probabilistic model to extract echoes bouncing off surrounding objects, which allows it to mitigate multipath effects for robust onset detection and thereby achieve accurate range estimation. BatMapper reports $1-2$ cm estimation errors with ranges up to 4 m. With the assistance of an on-board inertial measurement unit (IMU) including a gyroscope and an accelerometer, its 80-percentile errors are less than 30 cm in geometric floor reconstruction. SAMS [32] is another acoustic radar solution that improves upon BatMapper. Different from the correlation method adopted in BatMapper, SAMS utilizes chirp mixing, a finer temporal feature extraction method, which can circumvent insufficient sampling rates for better temporal resolution. SAMS reports a median error of 30 cm and a 90-percentile error of 1 m in map reconstruction.

Both ranging and acoustic radars only provide one-dimensional distance information. To obtain position information in 2D or 3D coordinates, one needs to resort to trilateration techniques as discussed next.

5.2 Localization

Localization is a key enabler for location-based service (LBS) such as navigation, asset management, augmented and virtual reality, etc. Despite tremendous research efforts on localization in GPS-denied environments such as indoor, underground, and urban canyons, many existing solutions either require expensive dedicated infrastructures [2, 43, 48] or rely on cumbersome device-dependent kernel hacking [4, 13, 14, 36], prohibiting their practical deployment. Among existing cutting-edge indoor localization approaches, acoustic-based systems attract much interests since they can achieve submeter-level localization accuracy with relatively low infrastructure costs and deployment efforts. The underlying building blocks for these localization systems are the timing or angle estimation methods discussed in Sects. 3.3 and 3.4. The common evaluation metric for localization systems is RMS error. In this section, we break down acoustic-enabled localization solutions in two categories, namely, *infrastructure-based* and *infrastructure-free* solutions.

5.2.1 *Infrastructure-Based Acoustic Localization*

Infrastructure-based schemes typically deploy low-cost, power-efficient distributed acoustic anchors in target areas, with their locations predetermined. Each anchor is equipped with acoustic transceivers and may include wireless modules to enable communication either among the anchors or with a remote server.

As shown in Fig. 5.1a, two primary configurations are possible: anchors as transmitters or anchors as receivers. In the first configuration, anchors periodically transmit acoustic beacons. When these transmitted signals are detected by a target device, associated information such as angles and timestamps (ToA or TDoA) is extracted. The target's location is then estimated using trilateration or more advanced optimization techniques. This approach offers advantages in scalability, as the number of beacon transmissions is independent of the number of targets in the area. It also enables low power consumption on target devices, which only need to passively activate their microphones to listen. However, this setting often requires more complex localization algorithms, and anchor transmissions may need to be synchronized and coordinated by a server. In contrast, when anchors act as receivers, they determine the target's location based on AoA estimations from multiple anchors upon receiving transmissions from the target device. While the latter approach is conceptually simpler, it scales poorly with the number of actively transmitting targets, making it more suitable for passive sensing scenarios, such as localizing sounds from humans or animals using an array of microphones.

Next, we examine synchronous schemes, where anchors transmit either concurrently or according to a predefined schedule. We then explore asynchronous approaches, in which anchor transmissions occur without coordination. Finally, we

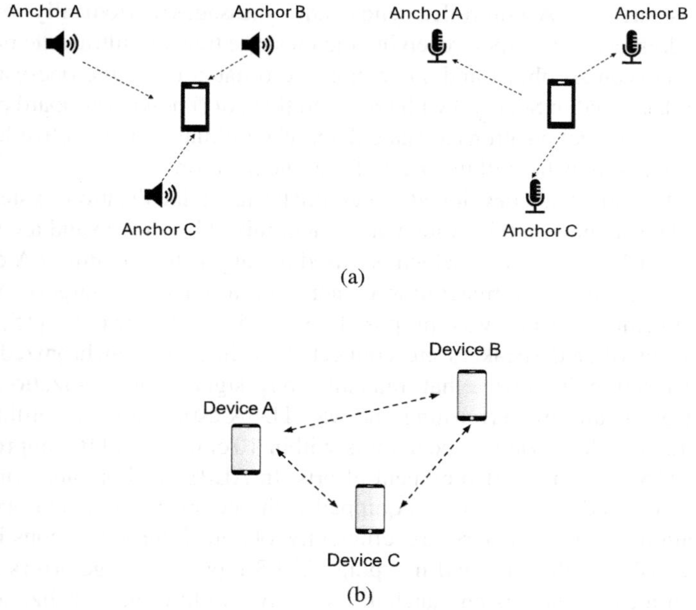

Fig. 5.1 Illustration of acoustic locations with and without infrastructure. (**a**) Infrastructure-based. (**b**) Infrastructure-free

Table 5.2 Comparison of infrastructure-based acoustic localization systems

Method	Waveform	Technique	Sync?	Accuracy
Active Bat [40]	Sinusoidal pulse	ToA (one-way)	Yes	dm-level
Guoguo [19]	Gaussian duplet pulse	ToA (one-way)	Yes	cm-level
[15]	Chirp	TDoA (one-way)	Yes	cm-level
ALPS [16]	Chirp	TDoA (one-way)	Yes	dm-level
UPS+ [17]	Ultrasonic chirp, tone	ToA (one-way)	Yes	cm-level
ARABIS [39]	Chirp	TDoA (two-way)	No	cm-level
AALTS [3]	Chirp, pure tone	TDoA (two-way)	No	dm-level
[8]	Exponential chirp	TDoA	No	Mean AoA 0.44°
RAILS [9]	Exponential chirp	AoA	No	cm-level
[6]	Walsh-Hadamard seq.	TDoA	No	AoA 0.4°–1.6°

discuss methods that rely on receiving-only anchors. A comparative summary of these approaches is provided in Table 5.2.

Synchronized Transmitting Anchors

Developed in 1997, Active Bat from AT&T Cambridge lab is arguably the first acoustic localization system using synchronized transmitting anchors [40]. In the system, each object to be localized is equipped with a 418 MHz radio transceiver. A matrix of receiver elements is mounted on the ceiling of target environments serving as anchors. A controller sends radio messages periodically with object device addresses, which upon receiving the message transmit ultrasonic pulses. The distance between an object and an anchor is estimated using the one-way sensing approach described in Sect. 3.3 with considerations of (known) on-board processing delay. Finally, object positions are calculated via multilateration. Active Bat reports 95% of raw readings lie within 14 cm of the true position.

In [19], Liu et al. developed a centimeter-level localization system named Guoguo. The anchors in this system are synchronized by Zigbee and are scheduled to transmit orthogonal codes, which are used by targets to perform ToA estimation via one-way sensing. Multilateration is then used to locate the targets. A speaker-only localization system was proposed by Lazik and Rowe in [15]. In their approach, distributed speakers are connected to different synchronized channels of an advanced audio device that transmits chirp signals for localization. A target locates itself locally by performing one-way TDoA estimation. According to [15], its 95-percentile localization accuracy is within 10 cm. ALPS [16] improves upon the work [15] in reduced deployment efforts. In ALPS, anchors are synchronized via Bluetooth, and each anchor is equipped with one microphone and one speaker. The locations of the anchors are efficiently obtained through acoustic-assisted simultaneously localization and mapping. ALPS reports average errors of 30 and 16.1 cm in locating targets and anchors. Recently, a ultrasonic localization system called UPS+ is presented in [17]. UPS+ leverages the nonlinearity of receiver microphones (details are presented in Sect. 2.1.1.3) to enable ultrasonic beacons to locate smart devices without ultrasonic sensors. Consequently, the audibility problem is eliminated. UPS+ achieves centimeter-level accuracy in localization.

Asynchronous Transmitting Anchors

The localization accuracy of the aforementioned work is highly dependent on clock synchronization accuracy, which in turn depends on network latency that is nonnegligible in a large-scale network. In contrast, asynchronous approaches can overcome such shortcomings. ARABIS [39] is an asynchronous acoustic localization system that utilizes two-way -ranging [31] to avoid the need for synchronization. In ARABIS, anchors transmit acoustic beacons periodically following a coarse TDMA schedule. Targets, as well as anchors, overhear the transmissions and record the corresponding timestamps. These timestamps are used to estimate TDoA information in locating a target. ARABIS reports a 95-percentile localization error of 7.4 cm. AALTS [3] improves upon ARABIS using a more robust onset detection approach to handle the near-far problem, hardware heterogeneity, and multipath effects. Specifically, it normalizes the current correlation value by the mean of a number of its preceding samples similar to Eq. (3.2) in Sect. 3.2. Additionally, a pseudo-orthogonal chirp spread spectrum modulation technique is proposed, which effectively doubles the transmission rate. AALTS achieves 90-percentile tracking errors of 0.49 m for mobile targets and a median of 0.12 m for stationary ones with only four anchor nodes.

Asynchronous Receiving-Only Anchors
To avoid the need for synchronization, in this line of work, individual anchors are equipped with microphone arrays to estimate AoAs from target devices' transmissions using techniques discussed in Sect. 3.4. Once the AoAs are obtained, the least-squares optimization is formulated to estimate the target position given the positions and orientations of each anchor.

In [8], Gabbrielli et al. propose a new robust, stable, accurate, and efficient time-delay estimation scheme for AoA using ultrasonic systems. The system setup consists of anchors (five microphones arranged on a pentagon of length $d_M = 8\,\text{cm}$) and ultrasonic speakers as tags. Each tag emits FMCW signals $x_{TC}(t) = A\cos(2\pi\phi(t))$, for $0 \leq t \leq T_c$ with exponentially modulated frequency, i.e., $\phi(t) = \left[\frac{f_0 T_c}{log(\frac{f_1}{f_0})}\left(\left(\frac{f_1}{f_0}\right)^{\frac{t}{T_c}} - 1\right)\right]$, where $f_0 = 18$ Khz, $f_1 = 30$ KHz. The received chirp signals go through match filtering to obtain CIR. Peak detection in the time domain signals from each microphone is used to estimate relative delays in the LoS paths. An echo suppression delay estimator (ESDE) is devised to suppress multipath effects. AoA is estimated from the TDoAs between the microphone elements (see Sect. 3.4.1). The solution achieves a mean absolute angle error of 0.44° and a median absolute angle error of 0.36° in a hall. Building upon ESDE for AoA estimations, the authors further developed, RAILS, an ultrasonic indoor localization system [9]. It reports an average 10-cm 3-D positioning accuracy in a warehouse-like scenario, covering an area of 40 m^2 with a maximum source-receiver range of 10 m.

In [6], a 8×8 microphone array is constructed to estimate the AoA of piezoelectric speaker tags, which transmit band-limited ($B = 200$ Hz) Walsh-Hadamard sequences mixed to 20 kHz. The MUSIC algorithm is applied and results in spherical errors ranging from 0.4° to 1.6°.

5.2.2 *Infrastructure-Free Acoustic Localization*

Infrastructure-free localization systems do not require the deployment of custom-built infrastructure devices in target areas. However, they tend to achieve less competitive localization performance than infrastructure-based solutions (Table 5.3).

In [18], Liu et al. built a localization system utilizing acoustic and WiFi signals. It first estimates pairwise distances within a device group using the acoustic ranging technique in [31], forming spatial constraints. Each device in the group also uses

Table 5.3 Comparison of infrastructure-free acoustic localization systems

Method	Waveform	Additional signal	Technique	Sync?	Accuracy
[18]	Chirp	WiFi	ToA (two-way)	No	m-level
Centaur [28]	Chirp	WiFi	TDoA (two-way)	No	m-level
EchoTag [34]	Chirp	NA	Data-driven	N/A	cm-level

WiFi fingerprints to impose another location constraint. By combining the two, target locations can be determined. This scheme achieves an 80-percentile localization error of 1 m. Since computing spatial constraints requires multiple pairwise acoustic ranging measurements and is computationally intensive, this approach is primarily suited for static target localization. Centaur [28], similar to [18], is a joint optimization framework utilizing acoustic and WiFi signals and reports meter-level localization accuracy. In [28], a novel multipath mitigation algorithm is proposed to achieve robust onset detection. The key idea is to inspect signal changes in cross-correlation as opposed to absolute magnitudes considered in prior methods. EchoTag [34] is an acoustic fingerprinting localization system that can detect minor location changes. It associates different acoustic profiles with different positions, known as tags, to train a classification model. This model is then used for online tag detection, enabling context-aware applications. EchoTag reports an accuracy of 98% in distinguishing 11 tags at 1 cm resolution. However, EchoTag is highly sensitive to environmental dynamics and experiences performance degradation when changes occur without additional data collection.

Summary
Compared to localization solutions utilizing RF signals [11], visible light [41], or IMU data [1], acoustic-enabled localization techniques strike good trade-offs between costs and accuracy. The acoustic diffraction property makes it feasible to locate targets in the presence of small-scale random blockages. This puts less restriction on deployment. However, due to limited transmission ranges, many anchor nodes are often needed in infrastructure-based localization systems to cover a large indoor space. Infrastructure-less solutions on the other hand either require close-by nodes for pairwise ranging or labor -intensive location fingerprint collection beforehand. It is worth noting that in both lines of approaches, acoustic signals are utilized for both sensing and communication. Specifically, acoustic signals are modulated to encode identifiers of acoustic anchors or tags using AAC techniques discussed in Chap. 4. The identifier information allows to distinguish the acoustic sources.

5.3 Tracking

High-accuracy object tracking is important in many applications such as automated surveillance, traffic monitoring, and augmented/mixed reality [44]. Tracking is a well-investigated topic in computer vision (CV) [26, 42, 44]. However, CV techniques impose substantial computation costs and do not work well under no or low- light conditions. Acoustic tracking systems can overcome these limitations. Existing acoustic solutions can be categorized into *device-based tracking* and *device-free tracking*, depending on whether the target object is capable of emitting acoustic signals.

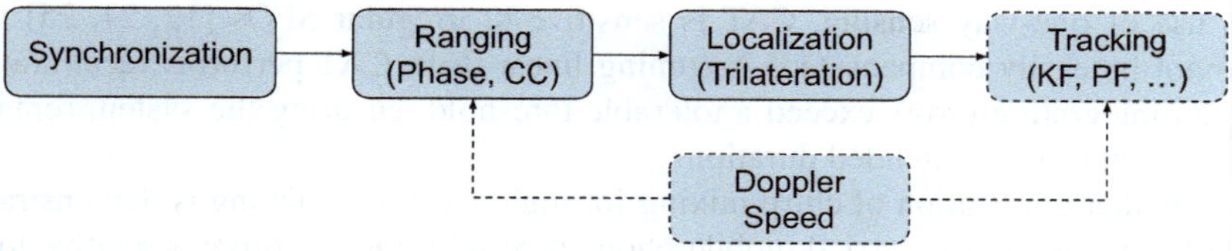

Fig. 5.2 Canonical device-based acoustic tracking pipeline. The dotted lines and blocks are optional. CC—cross-correlation, KF—Kalman filter, PF—particle filter

5.3.1 Device-Based Acoustic Tracking

To track acoustic emitting devices in motion, one can combine location estimation with movement information such as Doppler velocity estimated from received acoustic signals. The diagram of a typical pipeline for device-based acoustic tracking is given in Fig. 5.2.

AAMouse [45] tracks the location of a mobile phone or a smart watch and turns it into a mouse to control smart TV and laptops. In AAMouse, a controlled device (e.g., a TV) emits sound pulses at a few selected frequencies from multiple speakers, and a smart device (e.g., a phone) estimates the speed and distance traveled from the frequency shifts in its received signals. The radial velocity with respect to a speaker on the controlled device is determined by the estimated Doppler shift of the corresponding tone. The distance to the speaker is then updated by a displacement equal to the product of the radial velocity and time. From the distance measurements from multiple speakers, the location of the smart device is estimated through trilateration. To improve tracking accuracy, multiple tones are transmitted by each speaker. Maximal ratio combining is applied followed by Kalman filtering. AAMouse reports a median error of around 1.4 cm. Though the reported tracking performance is at centimeter-level, tracking errors can still accumulate over time, making it unsuitable for long-term operation. This limitation can be partially mitigated by incorporating map information and particle filtering, which can also aid in initial position estimation. However, this comes at the cost of increased computational complexity and longer response times.

CAT [20] improves upon AAMouse by adopting chirp mixing and boosts the tracking accuracy to a sub-centimeter level. Doppler shift is estimated using the same multifrequency tones as in AAMouse [45]. However, speakers on a controlled device also transmit FMCW signals for more precise ranging. A coarse-grained synchronization is done through cross-correlating the first received FMCW signals with a reference signal. This allows the synchronization of transmission and receiving intervals (e.g., 40 ms in [20]). Although the propagation delay is unknown, all subsequent onsets of received signals are assumed to experience the same delay. In other words, the changes in the distance between the transmitter and receiver can be determined. To convert the distance changes to absolute distances, a reference point with a known position is needed. This can be obtained by moving a phone between the two speakers of the controlled device, separated by a known distance. Due to its use of one-way sensing, CAT is sensitive to irregular SFOs [12, 24, 25] that cannot be easily compensated. Assuming linear drift, CAT performs recalibration when integration errors exceed a tolerable threshold, ensuring the system remains operational for an extended duration.

Another application of chirp mixing for high-accuracy tracking is demonstrated in [21], where a user-carried mobile phone precisely tracks a drone's relative location and controls its distance and orientation to enable automated video recording. In this chapter, the authors introduce several advanced signal processing modules. In particular, MUSIC is used to resolve multipath effects and enhance system robustness. A reciprocal filter is introduced to compensate the frequency selectivity problem (see the definition in Chap. 1) and thereby further enhance system stability. It is shown that with these techniques, a drone's location can be tracked within subcenter meter errors acoustically.

MilliSonic [37] is a device tracking solution for virtual reality (VR) applications using acoustic signals. The system consists of an FMCW chirp emitting device (e.g., a VR headset) and a four-microphone array. The two devices achieve synchronization through an initial calibration procedure. During this process, a user places the smartphone speaker in direct contact with the receiver's microphone, allowing the receiver to record a chirp signal for five seconds. 1D ranging is estimated from the phase change between the transmitted and received signals. 3D location is further computed from the intersection of all the resulting 1D positions between the target device and each microphone. MilliSonic achieves submillimeter 1D tracking accuracy using a single beacon, even in the presence of multipath interference. Multiple devices can be tracked simultaneously by employing a TDMA schedule.

5.3.2 Device-Free Acoustic Tracking

Device-free tracking allows tracking passive moving objects in an environment using reflected acoustic signals. Due to the significant attenuation of reflected signals, high-precision device-free acoustic tracking is often limited to short -range applications such as human-computer interactions.

FingerIO turns a mobile phone or a smart watch into an active sonar that is capable of tracking moving fingers for Around Device Interaction (ADI) [29]. Such a technology can extend the physical interaction boundaries and hence is particularly useful for small wearables. FingerIO achieves a median accuracy of 8 mm in 5.92 ms. It utilizes OFDM-modulated signals to estimate CSI between a hand and a smartphone periodically. In each estimation cycle, CSI is acquired through cross-correlation. Since only moving fingers can dynamically affect the channel between the speaker and microphone, their movements can be tracked by comparing consecutive channel frames. Static multipath reverberations remain the same across frames and can thus be removed. The proposed technique successfully addresses robust onset detection by extracting finger-movement-only CSI profiles and has inspired several follow-up work.

A multitone device-free gesture tracking system, LLAP, was proposed in [38]. It leverages coherent detection to extract the phases of acoustic echoes for finger localization and tracking. In LLAP, a mobile device actively transmits multiple tone carriers and decomposes finger-generated echoes via empirical mode decomposition (EMD) for later processing. It further uses the phase divergence of multiple carriers to coarsely locate the start position of a finger and track its displacement via phase shifts. LLAP reports a tracking accuracy of 3.5 mm for 1D hand movement and 4.57 mm for 2D drawing with less than 15 ms latency. However, both FingerIO and LLAP are sensitive to interference from close-by devices. To mitigate this issue, Strata [46] uses a coherent detector but applies a GSM training sequence modulated by binary phase shift keying (BPSK). Evaluation results demonstrate that it outperforms FingerIO and LLAP in all cases with an average tracking accuracy at 3 mm.

The aforementioned work considers micro-finger gesture tracking. For macro-body parts such as hand or the whole body, one often utilizes more powerful speakers to increase sensing range. The work in [22] shows that it is feasible to achieve room-level hand motion tracking with a customized platform. It employs an acoustic radar along with many advanced processing techniques including MIMO beamforming and deep learning for signal quality enhancement. The proposed system achieves 1.2–3.7 cm tracking errors within 4.5 m range and supports multiuser tracking. CovertBand [30] is an active sensing system for passive multiple object tracking. It builds on an active sonar with an enhanced speaker and uses the same parametric model as FingerIO [29] to track human body. CovertBand reports a median of 18 cm in tracking mobile targets. For static objects, it can achieve an accuracy of 8 cm with a distance up to 8 m in LOS conditions. A comparison of these tracking schemes is given in Table 5.4.

Summary

Device-based and device-free acoustic tracking can achieve centimeter-level accuracy due to the low propagation speed of acoustic waves in air, which allows for fine-grained time-of-flight and Doppler-based measurements. Depending on the use cases, various techniques have been developed to track objects of different sizes and motion characteristics, ranging from rigid movements of small handheld devices to full human body kinematics [33].

Table 5.4 Comparison of device-free gesture tracking systems

System	Waveform	Bandwidth (kHz)	Latency (ms)	Operation range (m)	Tracking error
FingerIO [29]	OFDM	18−20	5.92	<0.5	8 mm (2D)
LLAP [38]	Pure tones	17−23	≤15	0.5	3.5 mm (1D) 4.57 mm (2D)
Strata [46]	GSM sequence	18−22	12.5	0.5	3 mm
[22]	Chirp signal	18−20	40	4.5	1.2–3.7 cm
CovertBand [30]	OFDM	18−20	4.2	6	18 cm

Despite these advancements, existing methods face several limitations. High-noise environments remain a challenge, as environmental noise and interference can significantly degrade tracking performance. Additionally, coexistence issues arise when multiple acoustic systems operate in the same space, potentially causing interference, transmission collisions, or identifier switching. Future research should focus on robust noise mitigation techniques, adaptive signal processing, and efficient medium or spectrum sharing strategies to enhance the reliability and scalability of acoustic tracking systems.

References

1. N. Abhayasinghe, I. Murray, Human activity recognition using thigh angle derived from single thigh mounted IMU data, in *Proceedings of the IEEE IPIN* (2014), pp. 111–115
2. F. Adib et al., 3D tracking via body radio reflections, in *Proceedings of the 11st USENIX NSDI* (2014), pp. 317–329
3. C. Cai et al., Asynchronous acoustic localization and tracking for mobile targets. IEEE Int. Things J. **7**, 830–845 (2019)
4. Z. Chen et al., M^3: Multipath assisted wi-fi localization with a single access point. IEEE Trans. Mob. Comput. **20**(2), 588–602 (2021)
5. S. Counts, E. Fellheimer, Supporting social presence through lightweight photo sharing on and off the desktop, in *Proceedings of the 22nd ACM SIGCHI* (2004), pp. 599–606
6. G.K.J. Fischer et al., Evaluation of sparse acoustic array geometries for the application in indoor localization (2024). arXiv preprint arXiv:2406.09001
7. D. Frohlich et al., Requirements for photoware, in *Proceedings of the ACM CSCW* (2002), pp. 166–175
8. A. Gabbrielli et al., An echo suppression delay estimator for angle-of-arrival ultrasonic indoor localization. IEEE Trans. Instrument. Measure. **70**, 1–12 (2021)
9. A. Gabbrielli et al., RAILS: 3-D real-time angle of arrival ultrasonic indoor localization system. IEEE Trans. Instrument. Measure. **72**, 1–15 (2022)
10. D. Graham et al., A software-based sonar ranging sensor for smart phones. IEEE Int. Things J. **2**(6), 479–489 (2015)
11. H. Jiang et al., Smart home based on WiFi sensing: a survey. IEEE Access **6**, 13317–13325 (2018)
12. K. Kinoshita, T. Nakatani, Microphone-location dependent mask estimation for BSS using spatially distributed asynchronous microphones, in *2013 International Symposium on Intelligent Signal Processing and Communication Systems* (2013), pp. 326–331
13. M. Kotaru et al., Spotfi: Decimeter level localization using WiFi, in *Proceedings of the 2015 ACM Conference on Special Interest Group on Data Communication (SIGCOMM)*. SigComm '2015 (2015)
14. S. Kumar et al., Accurate indoor localization with zero start-up cost, in *Proceedings of the 20th Annual International Conference on Mobile Computing and Networking (MobiCom)*. MobiCom '2014 (2014)
15. P. Lazik, A. Rowe, Indoor pseudo-ranging of mobile devices using ultrasonic chirps, in *Proceedings of the 10th ACM Conference on Embedded Network Sensor Systems*. SenSys '12. (2012)
16. P. Lazik et al., ALPS: A bluetooth and ultrasound platform for mapping and localization, in *Proceedings of the 13th ACM Conference on Embedded Networked Sensor Systems*. SenSys '15. (2015)

17. Q. Lin, Z. An, L. Yang., Rebooting ultrasonic positioning systems for ultrasound-incapable smart devices, in *Proceedings. of the 25th ACM MobiCom* (2019), pp. 1–16
18. H. Liu et al., Push the limit of WiFi based localization for smart-phones, in *Proceedings of the 19th ACM MobiCom* (2012), pp. 305–316
19. K. Liu, X. Liu, X. Li, Guoguo: Enabling fine-grained indoor localization via smartphone, in *Proceeding of the 11th Annual International Conference on Mobile Systems, Applications, and Services*. MobiSys '13 (2013)
20. W. Mao, J. He, L. Qiu, CAT: High-precision acoustic motion tracking, in *Proceedings of the 22nd Annual International Conference on Mobile Computing and Networking* (2016), pp. 69–81
21. W. Mao et al., Indoor follow me drone, in *Proceedings of the 15th ACM MobiSys* (2017), pp. 345–358
22. W. Mao et al., RNN-based room scale hand motion tracking, in *Proceedings of the 25th ACM MobiCom* (2019), pp. 1–16
23. W. Mao et al., DeepRange: Acoustic ranging via deep learning, in *Proceedings of ACM Ubicomp* (2020), pp. 1–21
24. S. Miyabe, N. Ono, S. Makino, Blind compensation of inter-channel sampling frequency mismatch with maximum likelihood estimation in STFT domain, in *Proceedings of the IEEE International Conference on Acoustics, Speech and Signal Processing* (2013), pp. 674–678
25. S. Miyabe, N. Ono, S. Makino, Optimizing frame analysis with non-integrer shift for sampling mismatch compensation of long recording, in *Proceedings of IEEE Workshop on Applications of Signal Processing to Audio and Acoustics* (2013), pp. 1–4
26. S. Murray, Real-time multiple object tracking - A study on the importance of speed, in *CoRR*abs/1709.03572 (2017). arXiv: 1709.03572. http://arxiv.org/abs/1709.03572
27. T. Nadeem, L. Ji, Location aware IEEE 802.11 for spatial reuse enhancement. IEEE Trans. Mob. Comput. **6**(10), 1171–1184 (2007)
28. R. Nandakumar, K.K. Chintalapudi, V.N. Padmanabhan, Centaur: Locating devices in an office environment, in *Proceedings of the 18th ACM MobiCom* (2012), pp. 281–292
29. R. Nandakumar et al., FingerIO: Using active sonar for fine-grained finger tracking, in *Proceedings of the 34th ACM SIGCHI* (2016), pp. 1515–1525
30. R. Nandakumar et al., CovertBand: Activity information leakage using music, in *Proceedings of the 19th ACM UbiComp* (2017), pp. 1–24
31. C. Peng et al., Beepbeep: A high accuracy acoustic ranging system using cots mobile devices, in *Proceedings of the 5th ACM SenSys* (2007), pp. 1–14
32. S. Pradhan et al., Smartphone-based acoustic indoor space mapping, in *Proceedings of the 20th ACM UbiComp* (2018), pp. 1–26
33. Y. Shibata et al., Listening human behavior: 3d human pose estimation with acoustic signals, in *Proceedings of the IEEE/CVF Conference on Computer Vision and Pattern Recognition* (2023), pp. 13323–13332
34. Y.-C. Tung, K.G. Shin, EchoTag: Accurate infrastructure-free indoor location tagging with smartphones, in *Proceedings of the 21st Annual International Conference on Mobile Computing and Networking*. MobiCom '15 (2015)
35. M. Uddin, T. Nadeem, RF-Beep: A light ranging scheme for smart devices, in *Proceedings of the IEEE PerCom* (2013), pp. 114–122
36. D. Vasisht, S. Kumar, D. Katabi, Decimeter-level local-ization with a single WiFi access point, in *Proceedings of the 13th USENIX NSDI* (2016), pp. 165–178
37. A. Wang, S. Gollakota, MilliSonic: Pushing the limits of acoustic motion tracking, in *Proceedings of the 37th ACM SIGCHI* (2019), pp. 1–11
38. W. Wang, A.X. Liu, K. Sun, Device-free gesture tracking using acoustic signals, in *Proceedings of the 22nd ACM MobiCom* (2016), pp. 82–94
39. Y.-T. Wang et al., ARABIS: An asynchronous acoustic indoor positioning system for mobile devices, in *2017 IEEE International Conference on Indoor Positioning and Indoor Navigation (IPIN 2017)* (2017)

40. A. Ward, A. Jones, A. Hopper, A new location technique for the active office. IEEE Personal Commun. **4**(5), pp. 42–47 (1997)
41. Y.-L. Wei et al., CELLI: Indoor positioning using polarized sweeping light beams, in *Proceedings of the 15th ACM MobiSys* (2017), pp. 136–147
42. Y. Xiang, A. Alahi, S. Savarese, Learning to track: Online multi-object tracking by decision making, in *Proceedings of the 15th IEEE ICCV* (2015), pp. 4705–4713
43. J. Xiong, K. Jamieson, ArrayTrack: A fine-grained indoor location system, in *Proceedings of the 10th USENIX Conference on Networked Systems Design and Implementation*. NSDI '13 (2013)
44. A. Yilmaz, O. Javed, M. Shah, Object tracking: a survey. ACM Comput. Surv. **38**(4), 13–es (2006)
45. S. Yun, Y.-C. Chen, L. Qiu, Turning a mobile device into a mouse in the air, in *Proceedings of the 13th Annual International Conference on Mobile Systems, Applications, and Services*. MobiSys '15 (2015)
46. S. Yun et al., Strata: Fine-grained acoustic-based device-free tracking, in *Proceedings of the 15th ACM MobiSys* (2017), pp. 15–28
47. Z. Zhang et al., SwordFight: Enabling a new class of phone-to-phone action games on commodity phones, in *Proceedings of the 10th International Conference on Mobile Systems, Applications, and Services (MobiSys)*. MobiSys '12 (2012)
48. C. Zhang et al., iLocScan: Harnessing multipath for simultaneous indoor source localization and space scanning, in *Proceedings of the 12th ACM SenSys* (2014), pp. 91–104
49. B. Zhou et al., BatMapper: Acoustic sensing based indoor floor plan construction using smartphones, in *Proceedings of the 15th ACM MobiSys* (2017), pp. 42–55

Chapter 6
Physiological and Behavioral Sensing

6.1 Vital Sign Estimation

In this section, we start by examining the modeling that underpins physiological sensing. We then outline the specific challenges that have emerged in this field, and finally, literary works are introduced.

6.1.1 Modeling

The fundamental concept behind physiological sensing via acoustics is that physiological signals, such as respiration and heartbeat, can cause variations in the acoustic path length between a speaker, a reflecting surface (typically the chest), and a microphone, as illustrated in Fig. 6.1. When an acoustic device is directed toward a person to monitor their vital signs, the rhythmic breathing of the subject (inhalation and exhalation) can modulate the path length of reflected acoustic waves due to the subtle movement of their chest. Similarly, the beating of the heart can also cause such displacement, although it is less pronounced. This displacement occurs at the millimeter level, with respiration causing approximately 2–10 mm of chest movement [33] and heartbeat resulting in a displacement of 0.2–0.5 mm [54]. Considering that the sensing resolution of commercially available acoustic devices is approximately $c/B = 8.5$ mm [62], the more sensitive phase information of the acoustic signal is frequently used to detect these minute displacements. Here, c represents the propagation speed of acoustic signals, and B is the available inaudible bandwidth, typically ranging from 18 to 22 kHz [3–5]. Irrespective of the type of signal used, whether acoustic or radio waves, the channel response is typically modeled as

$$h(t) = \sum_{k=1}^{K} \beta_k(t) e^{-j2\pi f_c \frac{2d_k(t)}{c}} + n(t), \tag{6.1}$$

R. Zheng, C. Cai, *Acoustic Sensing on Commodity Devices and its Applications*, Wireless Networks, https://doi.org/10.1007/978-3-031-96875-4_6

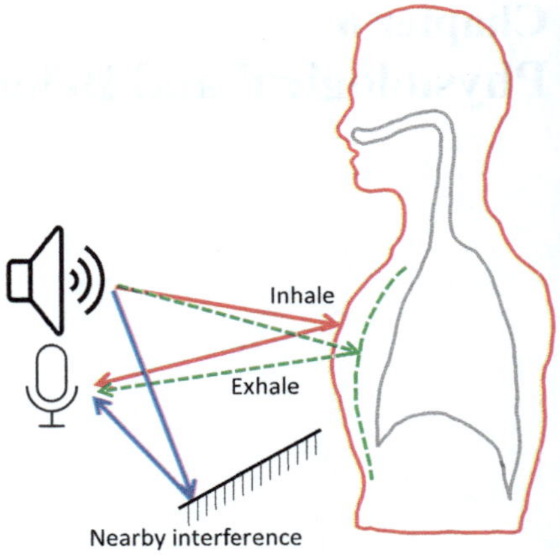

Fig. 6.1 Modeling for vital signs monitoring

where K denotes the number of reflections, $\beta_k(t)$ is a time-varying complex attenuation coefficient, f_c is the carrier frequency, d_k is the kth path length (from speaker to reflector and back to microphone), and $n(t)$ is the additive noise. In addition to vital signs, the displacement corresponding to a human reflector often contains body motions and hence is modeled as

$$d(t) = d_h(t) + d_r(t) + d_m(t), \tag{6.2}$$

where $d_h(t), d_r(t), d_m(t)$ denote displacements caused by heartbeat, respiration, and motion artifacts, respectively. As displacements caused by motion artifacts are often orders of magnitude greater than the vital signs, the sensing is often performed under static cases where $d_m(t) = 0$. And since respiration and heartbeat exhibit distinct fundamental frequency, $d_r(t)$ and $d_h(t)$ are often separated by appropriate filters [42, 43, 53, 72] or mode decomposition methods [30, 78, 81].

6.1.2 *Typical Processing Pipeline*

The enabling technique behind this vital sign sensing is actually radar, or sonar if using acoustic signals. A typical processing pipeline that leverages the aforementioned model in Eq. (6.1) to extract physiological signals is as follows: (i) designing an appropriate signal waveform for a sonar, (ii) extracting vital signs incurred displacement $d(t)$ via model $h(t)$ from interference, say motion artifacts or multipath effects, and (iii) separating mutual interfered vital signs leveraging distinct fundamental frequency. For illustration, we here use the most commonly used FMCW as an example.

When designing a suitable chirp waveform for vital signs monitoring, the most critical factor to consider is the bandwidth, as it dictates the achievable sensing resolution. However, the bandwidth on most acoustic devices is typically limited to 24 kHz, with the inaudible portion residing between 18 and 22 kHz [5]. To broaden the inaudible bandwidth, the researchers in [4, 6] suggest a phase whitening technique that modifies the chirp signal to be perceived as white noise. This approach can extend the bandwidth to 24 kHz, leading to a significant improvement in sensing resolution. However, it might necessitate a complex phase recovery process [4], which could potentially impact real-time performance. The chirp duration is another important factor to consider, as it influences both the maximum sensing range and the extent of multipath effects. Empirical evidence suggests that a duration of 0.04 s (at a sampling rate of 48 kHz) has proven to be effective.

A transmitted reference chirp $s(t) = \exp[-j(2\pi f_{\min} t + \pi k t^2)]$ would encounter propagation delay and complex attenuation before reaching a microphone:

$$r(t) = \exp[-j(2\pi f_{\min}(t - \Delta t) + \pi k(t - \Delta t)^2)]. \tag{6.3}$$

Here, $f_{\min}$ represents the starting frequency, $k = B/T$ is the chirp rate, and T indicates the duration of the chirp. To isolate the received (RX) frame depicted in Fig. 6.2a, a synchronization process is carried out using a matched filter, followed by synchronization to the transmitted (TX) frame, which typically exhibits a stronger signal strength in the received data. The received signals are then segmented into frames, each containing $f_s T$ samples (where f s is the sampling frequency), and processed sequentially to compute the corresponding channel responses $h(t)$, from which vital signs are derived.

Within each frame, a time delay Δt in the received signal RX corresponds to a proportional frequency shift Δf in the frequency domain relative to the transmitted signal (TX), as shown in Fig. 6.2a. Consequently, the process of inferring the channel response $h(t)$ from the received signal $r(t)$ within each frame is often performed in the frequency domain using a dechirp operation. This dechirp operation involves a mixing operation, denoted by $\odot$, between the received signal

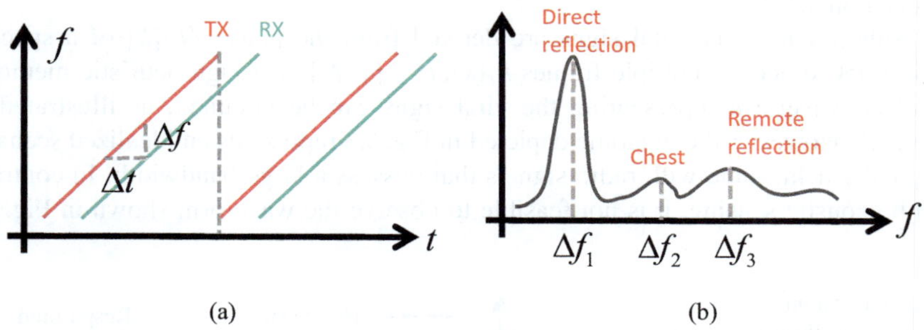

Fig. 6.2 Waveforms in the signal processing pipeline. (**a**) Transmitted and received signals. (**b**) Mixed signals

$r(t)$ and the transmitted chirp signal $s(t)$:

$$m(t) = r(t) \odot s(t) = \exp\left[-j(2\pi \Delta t \frac{B}{T} t + 2\pi f_{\min} \Delta t - \pi \frac{B}{T} \Delta t^2) + (\text{high freq. terms})\right]. \tag{6.4}$$

In practical implementation, this mixing operation is realized as a point-wise multiplication because the segmented signals $r(t)$ and $s(t)$ are discrete. Following the mixing operation, a low-pass filter is applied to eliminate the high-frequency components present in Eq. (6.4), and the resulting power spectrum is illustrated in Fig. 6.2b. In a static environment with minimal clutter, the power spectrum would typically exhibit two prominent peaks. These peaks correspond to the direct signal transmitted from the speaker to the microphone and the signal reflected from the chest, which travels from the speaker to the chest and then back to the microphone.

The peak in the power spectrum helps to locate the subject and lock the frequency index within which contains vital signs. Specifically:

$$\Delta f = B\Delta t / T. \tag{6.5}$$

And because $\Delta t = \frac{\Delta d}{c}$, Eq. (6.5) can be written as: $\Delta f = \frac{B\Delta d}{cT}$. Here, Δd refers to the distance between a sonar and a subject and therefore is often utilized to help lock the frequency index of the spectrum that contains vital signs. Since the spectrum at this time is descretized, we here denote it as $h[n]$. And because the frequency shift in this power spectrum infers the distance between a reflector and a sonar device, each frequency index in $h[n]$ is often termed as *range bin*, and the minimum distance a range bin denoted is often termed as *range resolution* Δr (determined by $\frac{c}{B}$). This range resolution can be boosted by padding zeros in the frequency domain while applying FFT to obtain $h[n]$. However, in most cases, the range resolution is configured to beyond the maximum displacement caused by vital sign signals ($\Delta r > \max(d_r(t)) > \max(d_h(t))$) so that one can observe them within a single locked bin $h[k]$.

Subsequently, the vital signs are derived from the phase $\angle h_i[k]$ of a specific bin, tracked across multiple frames $h_i[k], i \in [1, N]$. Through heuristic methods, a clear waveform representing the vital signs can be obtained, as illustrated in Fig. 6.3. However, the outcome depicted in Fig. 6.3 represents an idealized scenario typically achievable with radio signals that possess a large bandwidth. In contrast, with acoustic sensing, it is not feasible to observe the waveform shown in Fig. 6.3

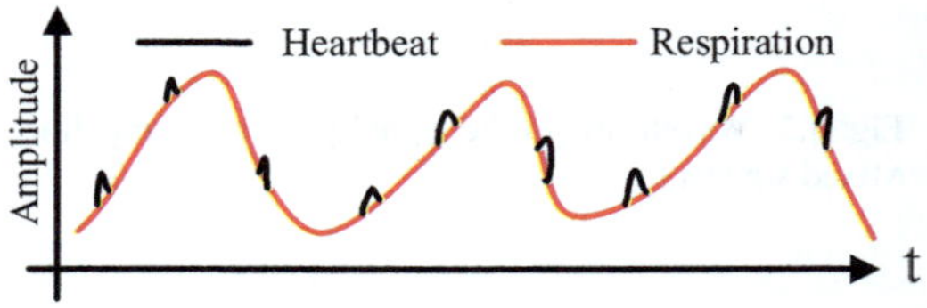

Fig. 6.3 An intuitive example of clean waveform that contains vital signs

if the range resolution Δr is set to be larger than the maximum displacement caused by vital signs, even when utilizing the entire available bandwidth of 24 kHz. Consequently, the vital signs often need to be extracted individually using different range resolutions rather than simultaneously. For example, to measure heartbeat signals, the range resolution would be configured (by zero padding) to be just above the maximum displacement expected from heartbeat. The rationale behind this is that a given displacement will result in a more significant phase change when the range resolution is finer. However, the phase of the tracked kth bin in $h[n]$ might not be sufficient to fully capture the subtle details of the heartbeat because respiration can cause the peak corresponding to the subject's chest in Fig. 6.2b to shift over time. One heuristic approach is to have the subject hold their breath during heartbeat measurements. Another intuitive solution involves combining the phase information from several range bins adjacent to the kth bin to reconstruct the heartbeat signals.

6.1.3 Challenges for Robust Physiological Sensing

In this section, we discuss common challenges for robust physiological sensing.

Multipath Effect

The multipath effect can introduce two primary challenges in this specific vital signs monitoring application. Firstly, it can obscure the accurate identification of the peak corresponding to the human subject within $h[n]$. As illustrated in Fig. 6.2b, reflections from distant objects can sometimes result in peaks of greater magnitude than that of the nearby chest, particularly if multiple reflections combine constructively. Furthermore, if a reflecting object is situated close to the subject, the resulting overlap of peaks can introduce significant noise into the phase within the respective range bin. Secondly, the initial phase $\angle h[i], i \in [1, N]$ of each carrier in $h[n]$ can be randomized by the multipath phenomenon, as previously noted in [83]. Consequently, if the vital sign signals are dispersed across multiple range bins, attempting to combine the phases from these bins to recover the signals can lead to abrupt changes and, therefore, noticeable noise.

Motion Artifacts

Motion artifacts can introduce displacement that is not only unpredictable but also significantly larger than the minute movements caused by vital signs. This disparity often leads to abrupt changes in the range bins that correspond to the monitored individual. One critical impact is that the system's sampling rate may be insufficient to adequately capture these rapid range bin variations, which carry the vital sign information. Another major challenge is the difficulty in correctly identifying the chest reflection when confronted with interference from other moving body parts, such as hands or the head, making accurate detection almost impossible. Moreover, the frequency content of particular motion artifacts, such as body movement modulated by leg shaking or treadmill running, can coincide with the frequency range of vital signs, resulting in severe interference. Compounding this issue, the spatial overlap between motion artifacts and vital signs makes spatial filtering strategies like beamforming unusable. Therefore, to ensure accuracy, most proposed methods resort to simply discarding the time periods affected by motion artifacts [60].

Dispersive Effect

The dispersive nature of the acoustic channel [83] can further complicate the accurate identification of the peak corresponding to vital signs. Ideally, the range bin Δf_2, as depicted in Fig. 6.2b, should uniquely correspond to the carrier whose phase is modulated by vital signs, provided the range resolution is sufficient to resolve the maximum displacement caused by these signs. However, in practice, the bins adjacent to Δf_2, whether to the left or right, might also contain vital sign signals within their phases, sometimes even with a superior signal-to-noise ratio (SNR). This phenomenon complicates the approach to extracting the displacement caused by vital signs. One might intuitively think that sampling the phases from multiple adjacent bins and then applying principal component analysis (PCA) could resolve this issue. However, the effectiveness of this method is typically limited to static scenarios.

Mutual Interference

The vital sign signals, namely respiration and heartbeat, are also spatially colocated and thus cannot be separated using spatial techniques, leading to mutual interference. A significant challenge arises from the substantial difference in amplitude between these two vital signs. Detecting the smaller signal, heartbeat, requires a sufficiently high sensitivity, perhaps achieved by enhancing the range resolution through interpolation. However, increased sensitivity inherently makes the system more susceptible to interference. When the sensitivity is high, the quantization errors resulting from the variations in range bins while sampling the larger respiration signal can introduce noise into the phase, potentially obscuring the faint heartbeat signals. Conversely, if the sensitivity is set low, sufficient to capture respiration but not optimized for heartbeat, there is a high probability that the heartbeat signal will be missed. One might suggest that increasing the number of samples and performing a long-sequence fast Fourier transform (FFT) could enhance the sensitivity to heartbeat, thus compensating for the reduced sensitivity. However, this approach can only provide statistical information, such as the heartbeat rate, over an extended measurement period, and would not be effective in quickly detecting irregular heartbeat patterns.

6.1.4 Existing Approaches Under Challenging Conditions

This section presents an overview of prior research in physiological sensing, with a specific focus on approaches designed to be robust against motion artifacts in vital sign monitoring. Current research in acoustic-based physiological sensing frequently focuses on extracting statistical information, such as heartbeat rate [53, 68, 78] or respiration rate [23, 62, 65, 66], primarily relying on spectral analysis and conducted under static conditions. The recovery of detailed waveforms, particularly for heartbeat, under adverse channel conditions such as motion artifacts, remains largely unexplored. Nevertheless, some preliminary efforts have been directed toward achieving motion-robust vital sign sensing. The underlying principles of these initial proposals are promising and could potentially stimulate further research in this area.

These works can be broadly classified into two categories: analytical signal processing methods and solutions leveraging deep learning techniques.

6.1.4.1 Analytical Signal Processing

The analytical signal processing methods leverage interpretable model to well describe the signal dynamics. The effectiveness of these methods heavily depends on the crafted models.

In the work by CORA [83], the researchers capitalize on the significant difference in amplitude between respiration and motion artifacts. To exploit this, they employ two types of signals, specifically frequency modulated continuous wave (FMCW) and orthogonal time-frequency space (OTFS) [21, 22], each with different sensitivities. The OTFS signal is designed to be sensitive solely to motion artifacts, denoted as $d_m(t)$, and insensitive to the displacement caused by respiration. Conversely, the FMCW signal is configured with high sensitivity to capture respiration, but it also registers substantial interference from motion artifacts, thus recording a composite signal $d_m(t) + d_r(t)$. The rationale behind this strategy is to exploit the inherent smoothness of human motion. This allows the less sensitive OTFS signals to effectively recover the detailed waveform of motion artifacts. Simultaneously, this configuration can also average out the minor variations caused by respiration when sampling the motion artifacts. The accurately recovered waveform of motion artifacts can then be used to compensate the FMCW signal, allowing for the isolation of the respiration signal through a subtraction process. This effectively separates the two colocated signals that were mutually interfering. An additional benefit of this approach is its robustness to significant multipath interference, which is attributed to the use of OTFS signals. This analytical solution enables the recovery of detailed respiration waveform information and allows for the timely detection of irregular breathing patterns. However, owing to bandwidth limitations, the heartbeat waveform cannot be easily extracted, although the heart rate might still be estimated using traditional spectral analysis techniques.

6.1.4.2 Deep-Learning-Enabled Techniques

The deep-learning (DL)-enabled techniques often involve in the second and third steps in the typical processing pipeline. The DL approaches though are not explainable but have demonstrated powerful capabilities. BreathListener [70] utilizes the audio capabilities of smartphones to estimate the detailed breathing waveform in driving scenarios. Their processing pipeline involves extracting the breathing pattern using energy spectrum density, employing background subtraction and ensemble empirical mode decomposition to reduce background noise, and transforming the extracted pattern to the Hilbert spectrum. Ultimately, a generative adversarial network is applied to this spectrum to reconstruct the fine-grained waveform. SpiroSonic [60], another approach, uses multiple pure tones to sense a combined signal of motion and respiration. Subsequently, neural network-based post-processing techniques are employed to reduce the impact of mild motion artifacts.

While these deep-learning-powered techniques demonstrate notable performance, they often require substantial amounts of training data, involve complex training procedures, and can suffer from issues related to generalizability.

6.2 Biometrics

Biometrics encompass the measurement and computational analysis of unique human characteristics and traits. Biometric authentication, also known as realistic authentication, serves as a method for identification and access control. This section will explore biometric acquisition techniques that utilize active sensing.

The fundamental principle of biometric sensing in this context is to identify unique patterns that can modulate actively transmitted acoustic signals. Common methods in this area involve the use of chirp signals, followed by the collection of resulting multipath profiles that are influenced by distinct biometric features, and ultimately, the application of classification techniques such as deep learning for authentication. Prior research has demonstrated that various biometric traits, such as facial patterns [82], the specific geometry of fingers during certain gestures [12], or handwritten signatures [13], can generate unique biometric signatures and, consequently, distinctive multipath profiles.

In [73], the authors present a customized resonant cavity designed to enhance the unique characteristics of fingerprints, offering an alternative approach to biometric authentication. Their method utilizes readily available and cost-effective smartphone cases, incorporating hollow-chambered mini-structures within them. These mini-structures exhibit both spatial and frequency diversity, enabling control over the propagation of actively transmitted structure-borne sound. When different users or fingers interact with the case, the unique biometric characteristics of the finger create distinct patterns that can be used for authentication. Specifically, the authors designed a series of chirp signals that sweep across the frequency range of 18 kHz–22 kHZ in 1 kHz steps. Each chirp sweep has a duration of 25 ms, which allows for the differentiation between structure-borne and airborne sound based on their differing propagation speeds. To prevent replay attacks, the transmitted signal patterns are randomized. For feature extraction, they utilize time domain features such as mean, standard deviation, maximum, minimum, range, kurtosis, and skewness, along with frequency domain features including spectral power and mel-frequency cepstral coefficients (MFCC), to create user-specific biometric templates.

In [67], researchers discovered that the deformation of the ear canal varies distinctly among individuals even when uttering the same words. Unlike traditional methods, this approach relies on the unique geometry of the ear canal, which is concealed within the skull, making it more resistant to spoofing attempts. Furthermore, this method allows for continuous authentication without requiring any active participation from the user. The researchers found that the geometry of the ear canal is influenced by an individual's anatomy and behavior, and the resulting patterns can be observed during speech. Human speech involves the movement of various articulators such as the jaw, tongue, and mouth to produce different phonemes. When the jaw moves, the temporomandibular joint also moves, leading to the expansion or compression of the ear canal wall. This phenomenon is termed ear canal dynamic motion (ECDM). The displacement and speed characteristics of an individual's ECDM during speech are unique and are reflected in acoustic profiles. Specifically, the authors first establish a mapping between different categories of ear canal deformation and the phonemes being spoken. Subsequently, they segment and label the captured audio during speech using the Munich Automatic Segmentation system. Finally, the channel response, also known as the frequency-amplitude response, is fed into an ensemble-learning-based classifier for authentication purposes.

6.3 Hearing and Auditory Perception

Hearing plays a crucial role in effective communication, social engagement, entertainment, and our overall quality of life. It allows us to connect with loved ones, navigate our environment safely, and enjoy experiences like music and conversations. Remarkably, with only binaural input, our auditory system can spatially locate sounds and selectively attend to relevant auditory inputs even amidst ambient noise and interference from other speakers, a phenomenon famously known as the "cocktail party problem." In contrast, unaddressed poor hearing health can lead to isolation, cognitive decline, and emotional challenges such as anxiety or depression. Regular assessment of hearing health and early detection of hearing problems are vital for enabling timely interventions, such as the use of hearing aids or other protective measures. As a result, there is a growing focus on creating techniques to measure ear acoustic responses and hearing parameters using widely available and user-friendly commodity devices.

6.3.1 Head-Related Transfer Function

Humans possess the ability to discern the spatial origin of sounds. Even in complex auditory scenes, such as a busy restaurant or a crowded stadium, individuals can selectively isolate and focus on specific sound sources [64]. This capacity for sound localization is due to the filtering effects imposed by the ears, head, and torso, which vary depending on the direction and frequency of the sound. These effects are described by the head-related transfer function (HRTF) [31]. HRTF defines how sounds originating from different locations in space are perceived by the ears; in essence, it represents a transfer function of the pathway between a sound source and the auditory system. Therefore, HRTF is a function of the angular coordinates of an incoming sound (e.g., azimuth and elevation in 3D interaural space) as well as frequency, and it is specifically defined for each ear.

Figure 6.4 illustrates an example of the head-related transfer function (HRTF) and its time domain counterpart, the head-related impulse response (HRIR), for both left and right ears. As demonstrated in Fig. 6.4b, numerous peaks and notches are observable in the frequency spectrum. When the sound source moves toward the top of the head, the frequencies at which these spectral notches occur tend to increase (Fig. 6.4c). The perception of a sound's elevation angle is linked to the spectral peaks and notches present above 5 kHz [24]. These spectral cues are dependent on the direction of the incoming sound signal as well as an individual's physical characteristics, such as the shape and size of their pinna, head, and torso. In contrast, interaural time difference (ITD) and interaural level difference (ILD) between the sounds arriving at the left and right ears are the primary cues for perceiving the lateral direction of a sound, and they are directly influenced by a person's HRTF [31].

Precise calculations of head-related transfer functions (HRTFs) for individuals are essential for facilitating binaural acoustic technologies like sound localiza-

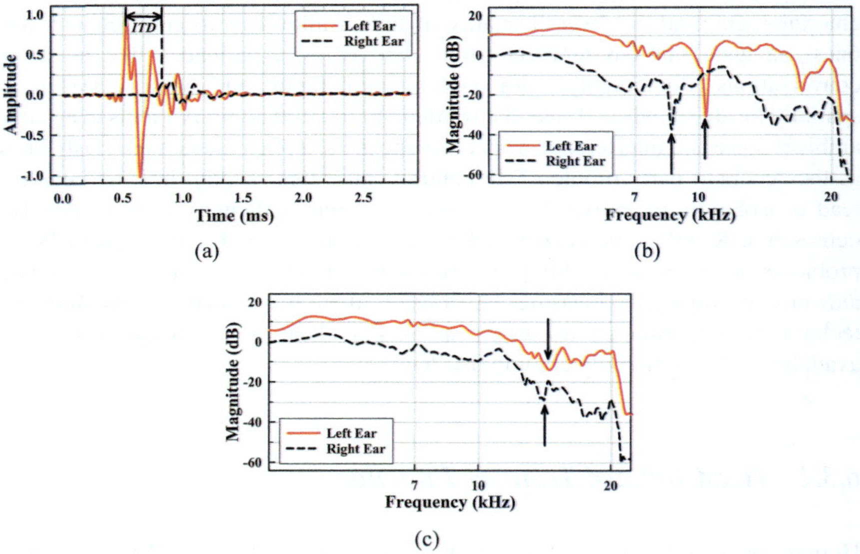

Fig. 6.4 Examples of HRIRs and HRTFs. (**a**) HRIR at azimuth = $-45°$ and elevation = $-5.76°$. (**b**) HRTF at azimuth = $-45°$ and elevation = $-5.7°$, and (**c**) HRTF at azimuth = 45° and elevation = 54.72° (some notches are marked with arrows). The notches appear at higher frequencies as the position of the sound moves toward the top of one's head

tion [18] and 3D sound spatialization, as well as binaural acoustic processing for hearing aids [16]. Given that HRTFs are highly personalized, employing another individual's HRTF or a nonspecific one will result in inaccuracies in sound localization and undesirable auditory experiences for listeners during audio playback. Current methods for personalizing HRTFs can be classified into four primary groups [19], which will be discussed in this section.

6.3.1.1 Direct Methods

The most straightforward approach to obtaining individualized HRTFs for a person involves conducting detailed acoustic measurements within an anechoic chamber [1]. This typically entails positioning loudspeakers around the subject in all relevant directions and using microphones placed at the entrance of their ear canals to record the impulse responses. The quantity of required loudspeakers can be reduced by arranging them at varying elevations along an arc and rotating this arc to perform measurements across different azimuth angles. This methodology necessitates specialized equipment and setups. The measurement process can be demanding for participants, often requiring them to remain still for extended periods. To expedite this process, the multiple exponential sweep method (MESM) is utilized in [44], where excitation signals are overlapped in time. However, this technique requires careful timing to avoid the superposition of different impulse responses. An alternative method is the reciprocal technique [86], where two small loudspeakers are placed inside the subject's ears, and microphones are positioned on an arc. While this reduces measurement time, it has its own drawbacks, as the in-ear loudspeakers cannot produce very loud sounds due to potential harm to the subject's hearing (leading to a low signal-to-noise ratio in the final measurements). More recently, some researchers have explored continuous measurement approaches, where instead of sequentially playing excitation waves from fixed locations, the sound sources move continuously through space. In [56], a study conducted in an anechoic room reported that a rotation speed of $3.8^\circ/s$ resulted in no perceivable differences for subjects compared to a step-wise measurement approach. In [55], rather than moving their entire body, subjects are instructed to move their head in various directions, with these head movements tracked by a motion capture system. Long measurement durations frequently lead to motion artifacts caused by subject movement during the data acquisition [26]. Some efforts have been made to mitigate this issue by providing visual feedback to the subjects [15] or by using mechanical supports to restrict their movements.

To alleviate the burden of time-consuming direct measurements, recent research has explored techniques for processing sparsely acquired data to generate more complete HRTF estimations. In [74], a user records their HRTFs at a limited number of 2D positions within a horizontal plane using a mobile phone. HRTF estimates are then derived by modeling sound diffraction around the head, integrating the principles of sound propagation with mobile phone inertial measurement unit (IMU) data to determine the phone's spatial location. Linear interpolation is employed to obtain HRTF estimations for locations where measurements were not taken. However, linear interpolation using nearby measurements is ineffective when the measurement points are widely spaced. Furthermore, as it can be difficult to perform measurements behind oneself, extrapolation based on frontal measurements often leads to significant errors. For instance, [74] reports that two out of five volunteers encountered difficulties in accurately measuring HRTFs at their back due to restrictions in arm movement. In [77], Zandi et al. propose a novel, computationally efficient algorithm for HRTF individualization. This algorithm, with the aid of a simple measurement procedure, enables users to individualize HRTFs at home using mobile devices.

Inspired by the observation that individual HRTFs exhibit both spatial and spectral sparsity, a *conditional variational autoencoder (CVAE)* is trained to learn their reduced-dimensional latent space representation from publicly available datasets. Given measurement data from a limited number of positions, the model can be adapted to generate personalized HRTFs for all directions (Fig. 6.5). The authors found that with as few as 60 measurement locations per user, which takes approximately 5 minutes to complete on a smartphone, accurate HRTF estimations can be achieved. The efficacy of this approach has also been confirmed through perceptual evaluations.

6.3.1.2 Simulation-Based Methods

Numerical simulation presents an alternative methodology for acquiring personalized HRTFs, particularly when direct measurement proves impractical. While simulation is a more time-consuming process than direct measurement, it eliminates

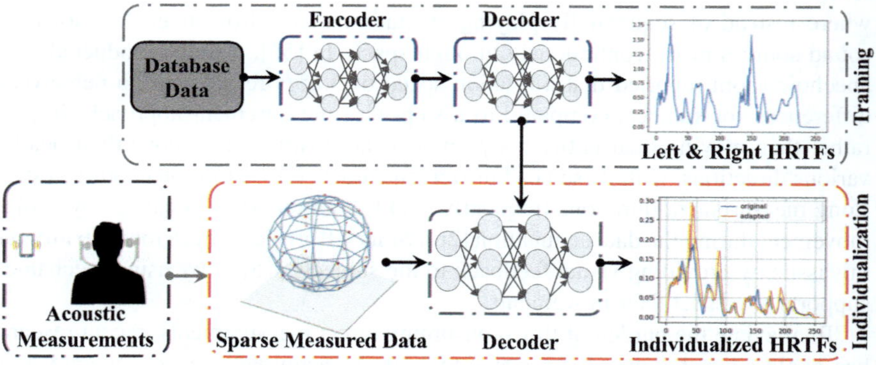

Fig. 6.5 System diagram of HRTF individualization from sparse measurements (reproduced from [77]). A conditional variational autoencoder is trained during the offline phase to embed HRTFs in a low-dimensional latent space. Sparse measurements from a new user are used to adapt the decoder parameters. HRTFs are generated from the decoder by sampling the latency space from a Gaussian distribution

the need for an anechoic chamber or an array of loudspeakers, thus making the approach more accessible. The primary requirement is a high-quality 3D model of the subject's head and ears [76]. The term "high quality" implies that the mesh is manifold, watertight, and without any holes. This mesh can then serve as the input for the simulation software. The HRTF is computed by employing numerical techniques to solve a system of partial differential equations that mathematically represent sound wave propagation [19]. For example, Mokhtari et al. utilized a magnetic resonance imaging (MRI) scanner to capture a 3D representation of the head's geometry and applied the finite difference time domain (FDTD) method to simulate acoustic wave propagation [49]. The boundary element method (BEM) is another widely used numerical technique for HRTF simulation. Ziegelwanger et al. implemented BEM in their open-source software package called Mesh2HRTF, which is utilized in this research to simulate individualized HRTFs [84]. Other notable contributions employing BEM include the work of Katz [34] and Zhang et al. [79].

Numerical simulation is emerging as a compelling method for acquiring individualized HRTFs, particularly due to its accessibility when paired with relatively inexpensive scanning technologies, such as Apple's FaceID depth sensor [58]. Furthermore, simulation offers enhanced accuracy compared to some previously discussed methods because the mesh input comprehensively captures the unique anatomical features of an individual, ranging from the cranium's form to the intricate folds of the outer ear (auricle/pinna). This contrasts with the generalized shapes obtained through anthropometric measurements or simplified spheroid-based mathematical models. The mesh-based depiction of the head's surface provides a highly detailed acoustic boundary for computational analysis, consequently yielding results that more closely approximate the actual HRTF [34].

However, obtaining a high-quality mesh can present difficulties due to obstructions that frequently occur during the scanning process [80]. These obstructions lead to incomplete data or other artifacts that often require manual correction through post-processing steps like filling holes and remeshing [2]. These methods demand considerable manual effort to prepare the mesh, lengthening the data acquisition phase and making the automated processing of multiple meshes impractical. Another challenge with simulation-based methods lies in their accuracy. The accuracy of acoustic simulations is contingent upon two primary factors: (1) how precisely the geometric and physical characteristics of surfaces (e.g., form, impedance) are captured and (2) the ability of the simulation tool to maintain precision across the relevant frequency spectrum. Regarding the former, studies have indicated that moisture levels and the presence of facial hair can influence the acoustic impedances of exposed skin, thereby affecting simulated HRTFs [17]. Regarding the latter, approximations made to accelerate the convergence of numerical methods can introduce inaccuracies in the outcomes. Consequently, simulated HRTFs tend to exhibit reasonable accuracy at low to medium frequencies, up to around 6 kHz, but show greater deviations at higher frequencies [17].

6.3.1.3 Indirect Methods Using Anthropometric Measurements

Since HRTFs are dependent on the listener's physical structure, numerous studies have attempted to indirectly estimate them using anthropometric data. These existing methods can be further categorized into three sub-approaches.

Adaptation Starting with a non-personalized HRTF, individualization can be achieved by applying scaling in the frequency domain [47]. The scaling factors for this process can be derived from measurements of the head and pinna [48]. Subjective assessments involving 9–11 participants demonstrated improved sound localization accuracy compared to using non-individualized HRTFs. Further enhancement can be realized by combining frequency scaling with spatial rotation to account for head inclination [45].

Nearest-Neighbor Selection These methods involve first selecting the most similar HRTF set from a database based on anthropometric measurements. The similarity between two individuals can be determined either directly from their morphological characteristics [85] or from features generated by a neural network [75]. Adaptation techniques from the previous category can then be applied for further refinement.

Regression The third category of approaches attempts to establish a mathematical or probabilistic relationship between anthropometric measurements and the defining features of HRTFs. Principal component analysis (PCA) is frequently employed to reduce the number of input and/or output variables [29]. In [29], a linear model is assumed, and the HRTFs for a new individual are predicted using their anthropometric parameters as input to the model. Subjective evaluations reported good agreement at various horizontal angles, but performance regarding vertical angles was not examined. Lee and Kim [38] and Chen et al. [11] build upon this work by modeling the two sets of parameters using a deep neural network and an autoencoder network, respectively.

All methods within this category share a common limitation with simulation-based methods: the requirement for precise anthropometric measurements, which are often challenging to obtain accurately.

6.3.1.4 Indirect Methods Based on Perceptual Feedback

In addition to using anthropometric parameters to identify closely matching individuals within a dataset, a fourth category of approaches relies on the subjective responses of target listeners. A reference sound, encompassing all audible frequencies (such as Gaussian noise or excerpts from music), is processed using selected HRTFs from a dataset and presented to the listener via headphones to generate virtual 3D audio experiences. The listener then evaluates these auditory presentations based on how accurately the perceived sound location aligns with the actual source location. Once the K-subjects most similar individuals in the dataset are identified, the listener's final HRTF can be determined through either selection or adaptation techniques.

In *selection* methods, the most perceptually similar non-individualized HRTFs from the dataset are directly utilized [35]. The reported time required for this tuning process varies from 15 to over 35 minutes. Conversely, *adaptation* methods employ frequency scaling, with the scaling factors adjusted based on the listener's perceptual feedback [27, 71]. In [71], the authors initially train a conditional variational autoencoder to generate HRTFs using a publicly available dataset. During the adaptation phase, user feedback is gathered and used to optimize personalized weighting of HRTFs from known subjects within the dataset. This approach reportedly yields significant improvements over non-individualized HRTFs for 18 out of 20 subjects. However, collecting user feedback necessitates at least 30 minutes.

Methods employing perceptual feedback are generally limited by lengthy calibration times and the inherent imperfections of human hearing (e.g., low resolution in perceiving elevation angles, difficulty in distinguishing sounds originating from directly in front or behind the body).

6.3.1.5 Summary

Creating techniques for simple and precise estimation of individualized HRTFs, without the need for specialized hardware, continues to be a vibrant area of investigation. However, the required degree of accuracy should be determined by the intended applications. Human hearing is quite robust to subtle variations in HRTFs and can adjust to substantial changes. For instance, while earplugs, earmolds, and electronic hearing aids can alter HRTFs and impact spatial hearing cues, studies have shown that training with feedback enables the human brain to adapt within 5–8 hours of the alteration [46]. Notably, these adaptations to the perception of auditory space can endure for several months, even without ongoing exposure to the learned cues. Consequently, when assessing HRTF individualization methods, it is crucial to take into account both quantifiable metrics, such as distortion levels, and qualitative metrics derived from perceptual evaluations.

6.3.2 Hearing Health

6.3.2.1 Hearing Screening

Affecting an estimated 430 million individuals, representing (5% of the world's population) moderate to severe hearing loss, is recognized as one of the most prevalent disabilities worldwide [69]. Even mild-to-moderate auditory impairments can have detrimental effects on communication abilities, personal well-being, and overall health and life quality. The seriousness of this problem is particularly pronounced in developing and underdeveloped countries, where roughly 80% of people with debilitating hearing loss live in low- and middle-income nations [69]. This imbalance can likely be linked to the lower socio-economic conditions prevalent in rural populations, often resulting in poor nutrition, insufficient health awareness, and inadequate access to medical care. These combined elements elevate the probability of developing hearing difficulties. Parallel patterns have been noted in industrial environments, where the use of protective equipment is often overlooked, and compliance with safety protocols is not strictly enforced.

A hearing impairment is classified as disabling when the hearing threshold exceeds 35 decibels (dB) in the better-hearing ear. The prevalence of hearing loss generally increases with age; notably, over 25% of individuals aged 60 and above experience disabling hearing loss. In clinical environments, hearing assessments often utilize otoacoustic emission (OAE) or auditory brainstem response (ABR) tests, which are conducted using specialized equipment [8]. However, these instruments are frequently costly and necessitate operation by trained technicians, thereby hindering widespread hearing screening in resource-limited countries. Consequently, numerous alternative approaches employing affordable and readily available devices, such as smartphones [8, 25, 39, 51] or wearables [7, 10, 14], have been suggested.

Standardized procedures exist for conducting hearing screenings using OtoAcoustic Emission (OAE) measurements. The primary methods employed are distortion product OAE (DPOAE) and transient-evoked OAE (TEOAE). A DPOAE response is generated by presenting two primary pure tones, labeled f_1 and f_2, to the inner ear. If the cochlea is functioning normally, a characteristic tone emerges at the frequency of $2f_1 - f_2$. An illustration of such a recording is provided in Fig. 6.6a, which displays the information in the frequency spectrum. To ascertain if the recorded signal at $2f_1 - f_2$ is a true OAE response or merely noise, it is crucial to evaluate the noise floor by averaging the signal levels in the adjacent frequency ranges. The signal-to-noise ratio (SNR) is then computed by comparing the magnitude of the signal at $2f_1 - f_2$ with the estimated noise level. A successful screening result ("pass") is only given when the SNR surpasses a defined threshold.

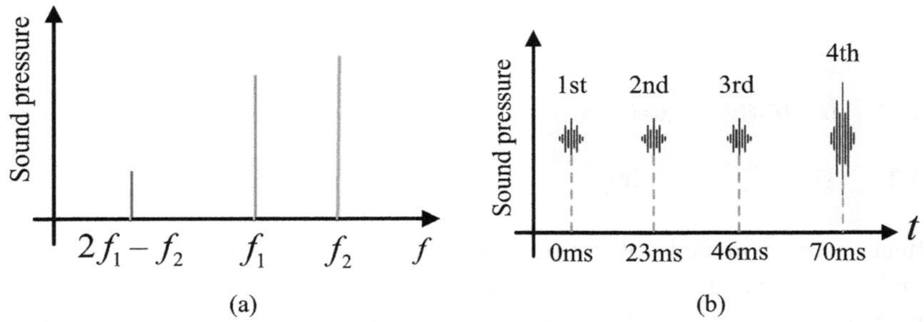

Fig. 6.6 OAE response function. (**a**) DPOAE. (**b**) TEOAE

Transient-evoked Otoacoustic Emissions (TEOAEs) are elicited using a click stimulus, typically with a duration of 100 μs and an intensity of 80 dB peSPL (peak equivalent sound pressure level). The inner ear's response to this wide-frequency signal can be detected in the ear canal within a few milliseconds. To distinguish the nonlinear OAEs from linear components and the stimulus echo, a nonlinear measurement technique involving a sequence of clicks is employed. This sequence consists of three clicks at a standard amplitude followed by a fourth click with three times the amplitude and reversed polarity. Figure 6.6b shows an example of a recorded response from this click sequence. By summing the responses from these four clicks, the linear elements are effectively eliminated, leaving only the nonlinear OAE and random noise. Similar to DPOAE measurements, it is essential to quantify both the background noise level and the signal strength. Multiple responses are recorded and compared, allowing the signal component to be estimated by averaging all responses, thereby reducing random noise. The noise level is estimated by calculating the standard deviation for each time point across all the recorded responses. The OAE signal arrives at the ear probe shortly after the stimulus presentation and is evaluated once the stimulus artifact has subsided. By calculating the root mean square (RMS) of the extracted signal and the noise within this specific time window, single values are obtained for both signal and noise levels. The SNR can then be computed and compared to a threshold in a manner analogous to DPOAE measurements to produce a result of "pass" or "refer." Due to the broadband nature of the stimulus, a single measurement is sufficient to assess the cochlear function across a wide range of frequencies.

To demonstrate the fundamental steps of an OAE test using readily available equipment, we will use DPOAE as an example. The necessary hardware for a DPOAE hearing screening includes an ear tip equipped with two speakers and a microphone. These two speakers emit two distinct pure tones at specific frequencies, denoted as f_1, f_2. The microphone then records the resulting sounds and uses fast Fourier transform (FFT) to determine if a harmonic frequency at $2f_1 - f_2$ is presented above the noise floor. To compensate for the frequency-dependent sensitivity of acoustic sensors and ambient noise, a series of frequency pairs are emitted. Furthermore, the sensitivity of the recording can be enhanced by increasing the window size of the FFT. A significant obstacle in performing this hearing screening on common devices lies in their non-optimized acoustic front ends, which can potentially obscure distorted OAE signals in the presence of background reverberations, leading to the incorrect identification of a healthy cochlea.

6.3.2.2 Ear Health Diagnosis

Recent years also witness a trend of developing ear examination techniques, say pneumatic otoscopy or tympanometry, using commodity devices [9, 32]. These commodity devices can act as pervasive, continuous ear condition monitoring devices that provide a long-term assessment of an individual's ear health.

Diagnosing ear health in this manner frequently necessitates headphones equipped with at least an internal microphone and a speaker. Alternatively, active noise cancelation headphones featuring a feedback microphone can also be utilized. The fundamental building blocks for this diagnostic process are the channel estimation techniques detailed in Chap. 3. Essentially, the speaker transmits modulated signals, and the microphone records the resulting reflections. Specific ear conditions, including a ruptured eardrum, earwax accumulation, and ear canal blockage, can alter the reflected acoustic profiles. By examining particular characteristics within these acoustic profiles, either through the use of physical models or deep learning methodologies, the ear's condition can be determined.

More specifically, otitis media can cause inflammation and lead to fluid buildup behind the eardrum. Earwax accumulation and blockage create more substantial reflective obstructions within the ear canal. A ruptured eardrum typically results in negative pressure, purulent effusion, and involvement of the middle ear cavity. All these phenomena can modify the sound wave transmission pathway and consequently generate distinctive acoustic reflection features [32]. Nevertheless, observing these altered acoustic characteristics, such as the spectrum obtained via FFT, resulting from these abnormal ear conditions can be challenging. To amplify these distinctions, the authors in [32] devised a set of spectral features as outlined in Table 6.1. In this table, *skewness* serves as a measure to evaluate the symmetry between the left and right sides of the center point. *Kurtosis* is a statistical method to characterize the probability distribution of the reflected spectrum. *Flatness* defines the extent to which the spectrum approximates a Euclidean space in a specific dimension. *Jitter* and *shimmer* are used to quantify the amplitude of eardrum vibration. The *crest factor* defines the sharpness of the correlation peak. The *mean cross* counts the number of local peaks that exceed the average signal level. By integrating these features into a deep learning classifier, the ear's status can then be readily identified.

Table 6.1 FFT features for abnormality detection [32]

Name	Calculation
Mean	$a(f) = \frac{1}{n}\sum_{i=1}^{n}(f_i)$
Deviation	$d(f) = \sqrt{\frac{1}{n}\sum_{i=1}^{n}(f_i - a(f))^2}$
Skewness	$s(f) = \frac{1}{n}\sum_{i=1}^{n}\left(\frac{f_i - a(f)}{d(f)}\right)^3$
Shimmer	$m(f) = \frac{\frac{1}{n-1}\sum_{i=1}^{n}\lvert f^{i+1} - f^{i}\rvert}{\frac{i}{n}\sum_{i=1}^{n} nf_i}$
Kurtosis	$k(f) = \frac{1}{n}\sum_{i=1}^{n}\left(\frac{f_i - a(f)}{d(f)}\right)^4$
Crest factor	$c(f) = 20\log\frac{\max_{i=1}^{n}\lvert f_i\rvert}{d(f)}$
Absorption	$p(f) = \frac{\sum_{i=1}^{n} f_i^2}{n}$
Auto-corr	$r(f) = \frac{\sum_{i=1}^{n-k}(f_i - \hat{f})(f_{i+k} - \hat{f})}{\sum_{i=1}^{n-k}\left(f_i - \hat{f}\right)^2}$
Mean-cross	$r_m(f) = count(f - \hat{f})$

6.4 Behavior Sensing

This section provides an overview of acoustic behavior sensing, encompassing areas such as gesture recognition [20, 57] and interactive sensing [36, 52, 63]. Compared to radio frequency (RF)-based methods, acoustic sensing solutions are capable of capturing more subtle over-the-air movements. For instance, while RF techniques are better suited for identifying large-scale body movements, acoustic approaches excel at detecting and tracking movements at the scale of fingers with higher accuracy. Furthermore, in structural-borne media like solid surfaces, any activity that alters the properties of the medium, such as its resonance frequency, can be detected using acoustic sensing. These distinctive characteristics pave the way for a broad spectrum of novel interactive sensing applications.

6.4.1 Over-the-Air Gesture Recognition

This section concentrates on gesture recognition facilitated by acoustic sensing. The fundamental concept is that hand movements generate unique acoustic echo profiles or Doppler signatures. Although these profiles can be influenced by factors like background noise or self-interference, they can be effectively isolated or improved using suitable signal processing methods, enabling accurate interpretation of the gestures.

In practice, acoustic gesture recognition employs acoustic radar principles. However, in contrast to vital sign sensing (Sect. 6.1) and motion tracking (Sect. 5.3.2), where monitoring changes in distance is paramount, the main goal here is to identify the patterns within the echoes generated by the gestures. Capturing the rapid changes in these echoes necessitates a sufficiently high sampling rate. This requirement renders long-duration waveforms, such as chirps, less appropriate for gesture recognition. Instead, spectrum-spread signals like Zadoff-Chu (ZC) sequences, Barker codes, and GSM training sequences are more frequently employed. The fundamental processing steps typically include: (i) extracting the echo profiles of the target gesture, (ii) recognizing basic gesture patterns using either deterministic or probabilistic methods, and (iii) performing high-level gesture recognition. To further elucidate the algorithmic procedures involved, we will now introduce two specific gesture recognition systems.

6.4.1.1 Doppler-Enabled Deterministic Gesture Recognition

The initial system employs a predictable method for recognizing gestures by leveraging the Doppler effect, as illustrated in Fig. 6.7a. In this approach, an acoustic emitter constantly sends out single-frequency sound waves toward a gesturing hand that is either approaching or receding along the line of sight.

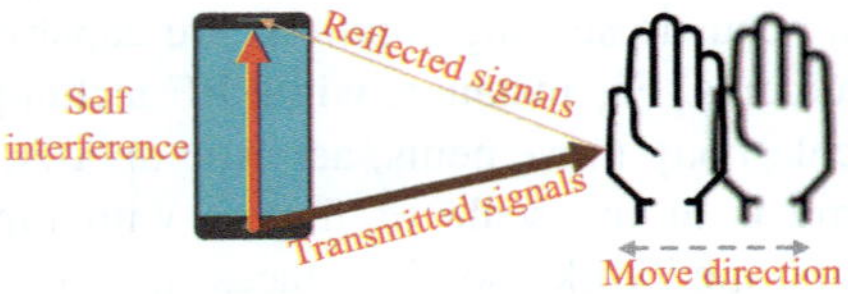

(a) Gesture recognition via Doppler effect.

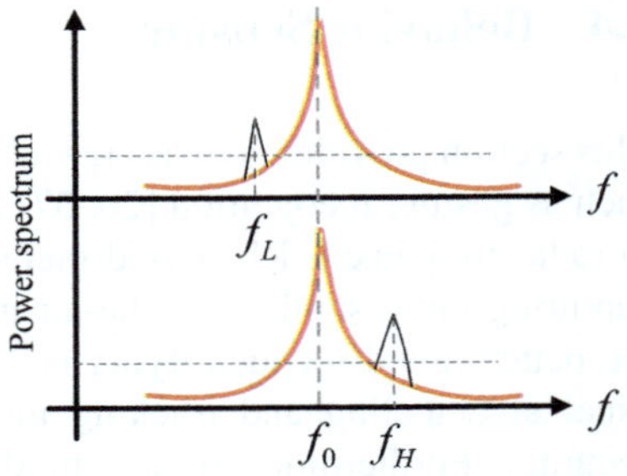

(b) Power spectrum of received signals.

Fig. 6.7 Gesture recognition using the Doppler effect (**a**) and the respective power spectrum when performed different gestures (**b**)

Fig. 6.8 The frequency response of comb-like filters

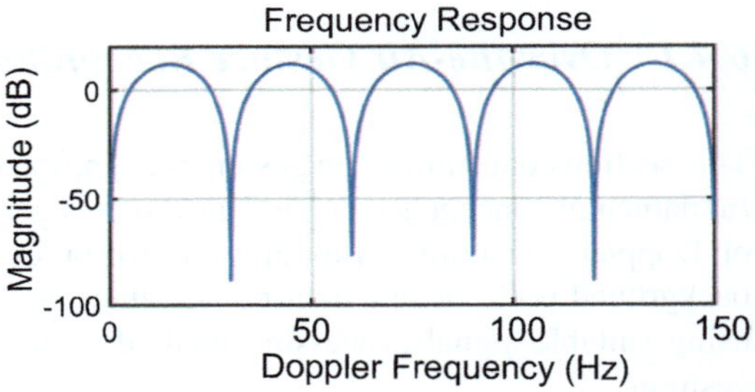

When the hand approaches the device, the returning sound signals exhibit an upward frequency change due to the Doppler effect; conversely, a receding hand causes a downward shift. This fundamental concept allows for the unambiguous identification of gestures by observing particular frequency-altered acoustic returns (f_L or f_H in Fig.6.7b) in comparison to the original transmitted frequency (f_0 in Fig.6.7b). However, the faint nature of these acoustic reflections often leads to their obscuration by self-interference, necessitating effective mitigation. To overcome this limitation, we propose a comprehensive framework for creating a resilient gesture recognition system.

Given the gesture recognition system's dependence on the Doppler effect, maximizing the Doppler resolution is crucial. This necessitates choosing the highest practically achievable frequency, denoted as f_0, for the emitted pure tone signal. Furthermore, employing zero padding during the fast Fourier transform (FFT) computation of the power spectrum can further improve resolution. To reduce self-interference, various filtering methods can be employed. Examples include the cascaded integrator-comb (CIC) filter [37, 50] and themoving target indicator (MTI) filter [40, 41]. These filters exhibit a frequency response characterized by multiple notches, as illustrated in Fig. 6.8. By aligning one of these frequency notches with the transmission frequency f_0, the impact of self-interference can be significantly lessened.

In a formal sense, let the frequency spectrum of the ith frame be represented by $\mathbf{h}_i$, where $\mathbf{h}_i \in \mathcal{R}^N$. If the frequency response of the filter applied is $\mathbf{g}$, then the resulting filtered spectrum

$$\mathbf{h'}_i$$

can be expressed as:

$$\mathbf{h}_i * \mathbf{g},$$

where $*$ signifies the convolution operation. Subsequently, a frequency window defined by the range $[f_0-L, f_0+L]$ is examined within $\mathbf{h'}_i$ to determine if the power spectrum within this band exceeds a predetermined threshold. Here, L represents half the window's width. By locating the instances where the power spectrum goes above this threshold, the fundamental gesture components can be identified. These basic gestures can then be combined to recognize more intricate interactions, such as double taps.

6.4.1.2 Channel-Based Probabilistic Gesture Recognition

Gesture recognition systems based on channel estimation draw inspiration from techniques used in wireless communication. The fundamental concept involves employing a designated signal, such as a ZC, GSM, or Barker code, to assess the dynamic characteristics of the communication channel between an acoustic emitter and a hand, while a gesture is being performed. The outcomes of this channel estimation process, specifically the channel state information (CSI), can exhibit clearly distinguishable patterns across various gestures, as depicted in Fig. 6.9.

The typical procedure for this approach involves an initial step of selecting an appropriate signal for channel estimation. The primary considerations during this selection are the system's ability to detect subtle changes (sensing sensitivity) and

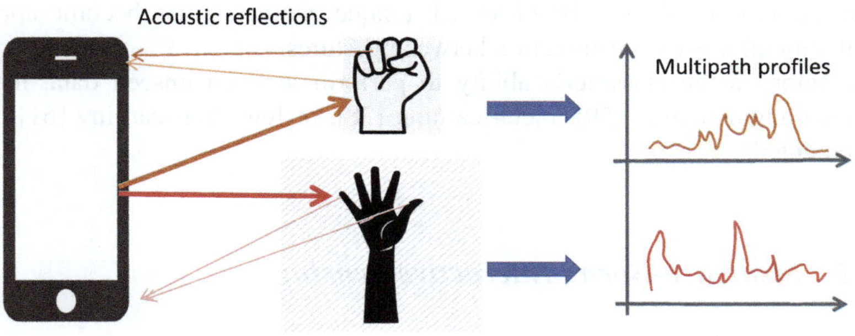

Fig. 6.9 Channel-based gesture recognition: Different gestures result in distinctive acoustic reflection profiles and hence signal matrix

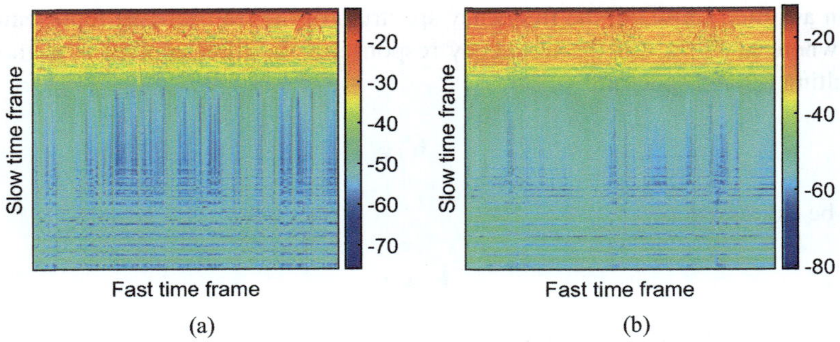

Fig. 6.10 Signal matrix when performing different gestures. (**a**) Signal matrix when drawing a circle. (**b**) Signal matrix when performing a push-pull gesture

the rate at which data frames are processed (frame rate). Sensitivity relates to the system's capacity to differentiate between multiple signal reflections arriving via different paths. Higher sensitivity ensures that the resulting CSI contains unique characteristics, like peaks that differentiate reflections from individual fingers. This improves the distinctiveness between different gestures, making them easier to classify. For comprehensive details on signal configuration, please refer to Chap. 2.

In the subsequent stage, the chosen signal is utilized for channel estimation. Let the estimated CSI for a given frame be represented by $\mathbf{h}$, where $\mathbf{h} \in \mathcal{R}^N$. By arranging multiple consecutive CSI frames within a defined time window, a signal matrix H is formed, where $H \in \mathcal{R}^{N \times M}$. To accentuate the features induced by the gesture and diminish the impact of strong, unchanging background reflections (static clutter), a differentiation operation can be applied along the temporal dimension of H.

Utilizing the resulting processed signal matrix, a connection between the recorded signals and the executed gestures can be determined. This established correlation can then be used to train a classification algorithm, for instance, a deep learning model, for real-time gesture recognition. Figure 6.10 provides illustrations of signal matrices generated using ZC signals for gestures like a double push and a circular motion. As can be observed, unique characteristics become apparent, facilitating effective discrimination between gestures.

To improve the classifier's ability to perform well on unseen data, methods like data augmentation [59], meta-learning [28], or few-shot learning [61] can be implemented.

6.4.2 Structure-Borne Interactive Sensing

This section introduces sensing systems that utilize the inherent structural acoustic characteristics of materials for interactive control. Specifically, these systems take advantage of acoustic dispersion and resonance phenomena, as detailed in Chap. 2. However, the processing steps involved in these systems differ considerably depending on the specific application, making it challenging to establish a universal processing framework. Due to the somewhat limited amount of research in this domain, we will present existing studies individually, based on their specific use case.

ForchPhone [63] is a touch force sensing system created for smartphones that do not have dedicated pressure sensors. It allows these smartphones to sense varying degrees of touch pressure using only software. The system conceptualizes the smartphone as a mass-spring system and sends out a modulated signal containing a wide range of frequencies from one part of the device, which is then picked up by a microphone on the opposite side. This arrangement causes the phone to vibrate and resonate at particular frequencies. When pressure is applied to the screen, the strength of these resonance frequencies changes. By examining these changes, the system can determine the level of touch pressure, enabling applications that respond to touch sensitivity.

Touch-and-Active [52] represents an interactive system for sensing both the location and force of touch, employing acoustic resonance. In this setup, a piezoelectric speaker and a microphone are affixed to the surface of an object. The system transmits a swept-frequency sinusoidal signal through the speaker, which is then recorded by the microphone, enabling the measurement of the object's acoustic response across a range of frequencies. When a user touches the surface, this action modifies the object's resonant characteristics, allowing the system to identify the interaction. If the object's form can be changed, each configuration displays a different resonance pattern, leading to a unique frequency response. Furthermore, the pressure exerted can cause a shift in the resonance frequency, enabling the system to simultaneously detect where the touch occurred and its intensity. This method facilitates precise sensing of interactions across various surface types.

References

1. R. Bomhardt, M. de la Fuente Klein, J. Fels, A high-resolution head-related transfer function and three-dimensional ear model database, in *Proceedings of Meetings on Acoustics* (2016)
2. H. Braren, J. Fels, A high-resolution individual 3D adult head and Torso model for HRTF simulation and validation: 3D data (2020)
3. C. Cai et al., Asynchronous acoustic localization and tracking for mobile targets. IEEE Int. Things J. **7**, 830–845 (2019)
4. C. Cai et al., AcuTe: Acoustic thermometer empowered by a single smartphone, in *Proceedings of the 18th ACM SenSys* (2020), pp. 28–41
5. C. Cai, R. Zheng, J. Luo, Ubiquitous acoustic sensing on commodity IoT devices: a survey. IEEE Commun. Surv. Tutor. **24**(1), 432–454 (2022)
6. C. Cai et al., Active acoustic sensing for "Hearing" temperature under acoustic interference.IEEE Trans. Mob. Comput. **22**(2), 661–673 (2023)

7. J. Chan, S. Gollakota, Inner-ear cochlea testing with earphones, in *Proceedings of the 20th Annual International Conference on Mobile Systems, Applications and Services*. MobiSys '22 (2022), pp. 609–610
8. J. Chan et al., An off-the-shelf otoacoustic-emission probe for hearing screening via a smartphone. Nat. Biomed. Eng. **6**, 1–11 (2022)
9. J. Chan et al., Performing tympanometry using smartphones, in Nat. Commun. Med. **57**, 397–408 (2022)
10. J. Chan et al., Wireless earbuds for low-cost hearing screening, in *Proceedings of the 21st Annual International Conference on Mobile Systems, Applications and Services* (2023), pp. 84–95
11. T.-Y. Chen, T.-H. Kuo, T.-S. Chi, Autoencoding HRTFS for DNN based HRTF personalization using anthropometric features, in *ICASSP 2019 - 2019 IEEE International Conference on Acoustics, Speech and Signal Processing (ICASSP)*. IEEE (2019)
12. H. Chen et al., Listen to your fingers: User authentication based on geometry biometrics of touch gesture. Proc. ACM Interact. Mob. Wearable Ubiquitous Technol. **4**(3), 1–23 (2020)
13. M. Chen et al., SilentSign: Device-free handwritten signature verification through acoustic sensing, in *2020 IEEE International Conference on Pervasive Computing and Communications (PerCom)* (2020), pp. 1–10
14. H.-L. Cheng et al., Objective signal analysis for investigating feasibility of active noise cancellation in hearing screening. Sensors **22**(19), 7329 (2022)
15. F. Denk et al., Controlling the head position during individual HRTF measurements and its effect on accuracy, in *Fortschritte der Akustik–DAGA, Kiel* (2017)
16. P. Derleth et al., Binaural signal processing in hearing aids, in *Seminars in Hearing*, vol. 42, no. 03 (Thieme Medical Publishers, New York, 2021), pp. 206–223
17. F. Di Giusto et al., Analysis of impedance effects on head-related transfer functions of 3D printed pinna and ear canal replicas, in *Forum Acusticum* (2023), pp. 1–8
18. A.M. El-Mohandes, N.H. Zandi, R. Zheng, DeepBSL: 3-D personalized deep binaural sound localization on earable devices. IEEE Int. Things J. **10**(21), 19004–19013 (2023)
19. C. Guezenoc, R. Seguier, HRTF individualization: A survey (2020). arXiv eprint, arXiv:2003.06183
20. S. Gupta et al., SoundWave: Using the doppler effect to sense gestures, in *Proceedings of the 30th ACM SIGCHI* (2012), pp. 1911–1914
21. R. Hadani, A. Monk, OTFS: A new generation of modulation addressing the challenges of 5G (2018). arXiv preprint arXiv:1802.02623
22. R. Hadani et al., Orthogonal time frequency space modulation, in *2017 IEEE Wireless Communications and Networking Conference (WCNC)* IEEE (2017), pp. 1–6
23. W. Haoran et al., RespTracker: Multi-user room-scale respiration tracking with commercial acoustic devices, in *Proceedings of the IEEE INFOCOM* (2021), pp. 1–10
24. J. Hebrank, D. Wright, Spectral cues used in the localization of sound sources on the median plane. J. Acoust. Soc. Am. **56**(6), 1829–1834 (1974)
25. N. Heitmann, T. Rosner, S. Chakraborty, Mass-deployable smartphone-based objective hearing screening with otoacoustic emissions, in *Proceedings of the 2021 International Conference on Multimodal Interaction* (2021), pp. 653–661
26. T. Hirahara et al., Head movement during head-related transfer function measurements. Acoust. Sci. Technol. **31**(2), 165–171 (2010)
27. J. Holzl, A Global Model for HRTF Individualization by Adjustment of Principal Component Weights. PhD thesis. Diploma Thesis, 2014
28. T. Hospedales et al., Meta-learning in neural networks: a survey. IEEE Trans. Pattern Analy. Mach. Intell. **44**(9), 5149–5169 (2022)
29. H. Hu et al., Head related transfer function personalization based on multiple regression analysis, in *2006 International Conference on Computational Intelligence and Security*. IEEE (2006)
30. W. Hu et al., Noncontact accurate measurement of cardiopulmonary activity using a compact quadrature doppler radar sensor. IEEE Trans. Biomed. Eng. **61**(3), 725–735 (2014)

31. K. Iida, *Head-related transfer function and acoustic virtual reality* (Springer, Singapore, 2019). ISBN: 978-98-11397-45-5
32. Y. Jin et al., EarHealth: An earphone-based acoustic otoscope for detection of multiple ear diseases in daily life, in *Proceedings of the 20th Annual International Conference on Mobile Systems, Applications and Services* (2022), pp. 397–408
33. H. Kaneko, J. Horie, Breathing movements of the chest and abdominal wall in healthy subjects. Respirat. Care **57**(9), 1442–1451 (2012)
34. B.F.G. Katz, Measurement and Calculation of Individual Head-Related Transfer Functions Using a Boundary Element Model Including the Measurement and Effect of Skin and Hair Impedance. Theses. The Pennsylvania State University (1998). https://hal.sorbonne-universite.fr/tel-02641309
35. B.F.G. Katz, G. Parseihian, Perceptually based head-related transfer function database optimization. J. Acoust. Soc. Am. **131**(2), EL99–EL105 (2012)
36. H. Kim et al., UbiTap: Leveraging acoustic dispersion for ubiquitous touch interface on solid surfaces, in *Proceedings of the 16th ACM SenSys* (2018), pp. 211–223
37. A.Y. Kwentus, Z. Jiang, A.N. Willson, Application of filter sharpening to cascaded integrator-comb decimation filters. IEEE Trans. Signal Process. **45**(2), 457–467 (1997)
38. G. Lee, H. Kim, Personalized HRTF modeling based on deep neural network using anthropometric measurements and images of the ear. Appl. Sci. **8**(11), 2180 (2018)
39. S.Y. Lee et al., Assessing the accuracy and reliability of application-based audiometry for hearing evaluation. Sci. Rep. **14**, 1–8 (2024)
40. N. Levanon, Radar, in *Encyclopedia of Physical Science and Technology (Third Edition)*, ed. by R.A. Meyers, 3rd edn. (Academic, New York, 2003), pp. 497–510. ISBN: 978-01-2-2274-10-7
41. H.-J. Li, Y.-W. Kiang, Radar and inverse scattering, in *The Electrical Engineering Handbook*, ed. by W.-K. Chen (Academic, New Yrok, 2005), pp. 671–690. ISBN: 978-01-21709-60-0
42. F. Lin et al., Cardiac scan: A non-contact and continuous heart-based user authentication system, in *Proceedings of the 23rd ACM MobiCom* (2017), pp. 315–328
43. J. Liu et al., Tracking vital signs during sleep leveraging off-the-shelf WiFi, in *Proceedings of the 16th ACM MobiHoc* (2015), pp. 267–276
44. P. Majdak, P. Balazs, B. Laback, Multiple exponential sweep method for fast measurement of head-related transfer functions. J. Audio Eng. Soc. **55**(7/8), 623–637 (2007)
45. K. Maki, S. Furukawa., Reducing individual differences in the external-ear transfer functions of the Mongolian gerbil. J. Acoust. Soc. Am. **118**(4), 2392–2404 (2005)
46. C. Mendonça, A review on auditory space adaptations to altered head-related cues. Front. Neurosci. **8**, 219 (2014)
47. J.C. Middlebrooks, Individual differences in external-ear transfer functions reduced by scaling in frequency. J. Acousti. Soc. Am. **106**(3), 1480–1492 (1999)
48. J.C. Middlebrooks, E.A. Macpherson, Z.A. Onsan, Psychophysical customization of directional transfer functions for virtual sound localization. J. Acoust. Soc. Am. **108**(9), 3088–3091 (2000)
49. P. Mokhtari et al., Computer simulation of HRTFs for personalization of 3D audio, in *2008 Second International Symposium on Universal Communication* (2008), pp. 435–440
50. G. Molnar, A. Dudarin, M. Vucic, Design and multiplierless realization of maximally flat sharpened-CIC compensators. IEEE Trans. Circ. Syst. II Express Briefs **65**(1), 51–55 (2018)
51. Y. Na et al., Smartphone-based hearing screening in noisy environments. Sensors **14**(6), 10346–10360 (2014)
52. M. Ono, B. Shizuki, J. Tanaka, Sensing touch force using active acoustic sensing, in *Proceedings of the Ninth International Conference on Tangible, Embedded, and Embodied Interaction*. TEI '15 (2015)
53. K. Qian et al., Acousticcardiogram: Monitoring heartbeats using acoustic signals on smart devices, in *Proceedings of the 37th IEEE INFOCOM* (2018), pp. 1574–1582
54. G. Ramachandran, M. Singh, Three-dimensional reconstruction of cardiac displacement patterns on the chest wall during the P, QRS and T-segments of the ECG by laser speckle inteferometry. Med. Biol. Eng. Comput. **27**, 525–530 (1989)

55. J. Reijniers et al., HRTF measurement by means of unsupervised head movements with respect to a single fixed speaker. IEEE Access **8**, 1–1 (2020)
56. J.-G. Richter, J. Fels, On the influence of continuous subject rotation during high-resolution head-related transfer function measurements. IEEE/ACM Trans. Audio Speech Lang. Process. **27**(4), 730–741 (2019)
57. W. Ruan et al., AudioGest: Enabling fine-grained hand gesture detection by decoding echo signal, in *Proceedings of the 18th ACM UbiComp* (2016), pp. 474–485
58. H.L. Rudy et al., Three-dimensional facial scanning at the fingertips of patients and surgeons: accuracy and precision testing of iPhone X three-dimensional scanner. Plast. Reconstr. Surgery **146**(6), 1407–1417 (2020)
59. C. Shorten, T.M. Khoshgoftaar, A survey on image data augmentation for deep learning. J. Big Data **6**(60), 1–48 (2019)
60. X. Song et al., SpiroSonic: Monitoring human lung function via acoustic sensing on commodity smartphones, in *Proceedings of the 26th ACM MobiCom* (2020), pp. 1–14
61. Y. Song et al., A comprehensive survey of few-shot learning: evolution, applications, challenges, and opportunities. ACM Comput. Surv. **55**(13s), 1–40 (2023)
62. Y. Su et al., Embracing distributed acoustic sensing in car cabin for children presence detection. Proc. ACM Interact. Mob Wearable Ubiquit. Technol. **8**(1), 1–28 (2024)
63. Y.-C. Tung, K.G. Shin, Expansion of human-phone interface by sensing structure-borne sound propagation, in *Proceedings of the 14th ACM MobiSys* (2016), pp. 277–289
64. D. Wang, *Computational Auditory Scene Analysis: Principles, Algorithms, and Applications* (Wiley-IEEE Press, Chichester, 2006). ISBN: 978-04-70043-38-7
65. T. Wang et al., C-FMCW based contactless respiration detection using acoustic signal. Proc. ACM Interact. Mob. Wearable Ubiquit. Technol. **1**(4), 1–20 (2018)
66. A. Wang, J.E. Sunshine, S. Gollakota, Contactless infant monitoring using white noise, in *Proceedings of the 25th ACM MobiCom* (2019), pp. 1–16
67. Z. Wang et al., EarDynamic: An ear canal deformation based continuous user authentication using in-ear wearables. Proc. ACM Interact. Mob. Wearable Ubiquit. Technol. **5**(1), 1–27 (2021)
68. L. Wang et al., LoEar: Push the range limit of acoustic sensing for vital sign monitoring. Proc. ACM Interact. Mob. Wearable Ubiquit. Technol. **6**(3), 1–24 (2022)
69. *World Health Organization: deafness and hearing loss*. https://www.who.int/news-room/fact-sheets/detail/deafness-and-hearing-loss (2017)
70. X. Xu et al., BreathListener: Fine-grained breathing monitoring in driving environments utilizing acoustic signals, in *Proceedings of the 17th ACM MobiSys* (2019), pp. 55–64
71. K. Yamamoto, T. Igarashi, Fully perceptual-based 3D spatial sound individualization with an adaptive variational autoencoder. ACM Trans. Graph. **36**(6), 1–13 (2017)
72. Z. Yang et al., Vital sign and sleep monitoring using millimeter wave. ACM Trans. Sensor Netw. **13**(2), 1–32 (2017)
73. Y. Yang et al., BioCase: Privacy protection via acoustic sensing of finger touches on smartphone case mini-structures, in *Proceedings of the 21st Annual International Conference on Mobile Systems, Applications and Services* (2023), pp. 397–409
74. Z. Yang, R. Roy Choudhury, Personalizing head related transfer functions for earables, in *Proceedings of the 2021 ACM SIGCOMM 2021 Conference* (2021), pp. 137–150
75. S.-N. Yao, T. Collins, C. Liang, Head-related transfer function selection using neural networks. Arch. Acoust. **42**(3), 365–373 (2017)
76. K. Young, T. Tew, G. Kearney, Boundary element method modelling of KEMAR for binaural rendering: Mesh production and validation, in *Interactive Audio Systems Symposium* (York, York, 2016). Visited on 11/18/2024
77. N.H. Zandi, A.M. El-Mohandes, R. Zheng, Individualizing head-related transfer functions for binaural acoustic applications, in *2022 21st ACM/IEEE International Conference on Information Processing in Sensor Networks (IPSN)* (2022), pp. 105–117
78. F. Zhang et al., Your smart speaker Can "Hear" Your Heartbeat!", in *Proceedings of the 29th ACM UbiComp*, vol. 4, no. 4 (2020), pp. 1–24

79. M. Zhang, J.-H. Wang, D.L. James, Personalized HRTF modeling using DNN-augmented BEM, in *ICASSP 2021 - 2021 IEEE International Conference on Acoustics, Speech and Signal Processing (ICASSP)* (2021), pp. 451–455
80. C. Zhang, H. Zhou, J. Duan, A method for identifying and repairing holes on the surface of unorganized point cloud. Measurement **210**, 112575 (2023). ISSN: 0263-2241. Visited on 11/18/2024
81. T. Zheng et al., V^2iFi: In-vehicle vital sign monitoring via compact RF sensing, in *Proceedings of the 22nd ACM UbiComp*(2020), pp. 1–27
82. B. Zhou et al., Robust human face authentication leveraging acoustic sensing on smartphones. IEEE Trans. Mob. Comput. **21**(8), 3009–3023 (2022)
83. J. Zhou et al., CORA: COntinuous respiration monitoring using analytical signal processing. IEEE Trans. Mob. Comput. **23**, 1–16 (2024)
84. H. Ziegelwanger, W. Kreuzer, P. Majdak, Mesh2HRTF: An open-source software package for the numerical calculation of head-related transfer functions, in *Proceedings of the 22nd International Congress on Sound and Vibration (ICSV)*. Florence (2015), p. 42583
85. D.N. Zotkin, R. Duraiswami, L.S. Davis, Customizable auditory displays, in *Proceedings of the International Conference on Auditory Display*. Georgia Institute of Technology (2002)
86. D.N. Zotkin et al., Fast head-related transfer function measurement via reciprocity. J. Acoust. Soc. Am. **120**(4), 2202–2215 (2006)

Appendix A
Selected Code Examples

Reference implementations are provided as supplement materials to this book to Elucidate the basic concepts and key algorithms discussed. They are organized into folders simplify navigation and usage.

A.1 Folder Structure

```
code/
|----[aac]/
|     |----chirp_up.m
|     |----chirp_down.m
|     |----CSS.m
|     |----HRCSS.m
|     |----OCDM.m
|     |----OCSS.m
|     |----OFDM.m
|     |----params.m
|     |----PreambleDetection.m
|     |----PreambleDetectionTriple.m
|     |----PreambleNormalization.m
|----[OFDM-OTFS]/
|     |----[Functions]/
|           |----Block_LMMSE_detector.m
|           |----chirp_up.m
|           |----convDecode.m
|           |----convEncode.m
|           |----evar.m
```


R. Zheng, C. Cai, *Acoustic Sensing on Commodity Devices and its Applications*, Wireless Networks, https://doi.org/10.1007/978-3-031-96875-4

```
|          |----Gen_DD_and_DT_channel_matrices.m
|          |----Gen_delay_time_channel_vectors.m
|          |----Gen_time_domain_channel.m
|          |----Generate_2D_data_grid.m
|          |----Generate_time_frequency_channel_ZP.m
|          |----MPA_detector.m
|          |----MRC_delay_time_detector.m
|          |----params.m
|          |----TF_single_tap_equalizer.m
|     |----[Gen_data]/
|          |----data_OFDM.mat
|          |----data_OTFS.mat
|          |----decoded_data.mat
|          |----decoded_data_raw.mat
|          |----transmit_data.mat
|          |----transmit_data_raw.mat
|          |----transmit_signal_OFDM.mat
|          |----transmit_signal_OTFS.mat
|     |----generator.m
|     |----params.m
|     |----receiver_OFDM.m
|     |----receiver_OTFS.m
|----[tdoa-localization]/
|     |----[anchor]/
|     |----[android_app]/
|     |----[images]/
|     |----[platform_cc_single_transmission_neon_ten_tones]/
|     |----[serve]/
|     |----README.md
|----[two-way-ranging-android]/
|----phase_modulation.m
|----waveform_example.m
```

A.2 Description of Content

- **[aac]**
  - Implements several AAC pipelines including CCS, OCSS, OFDM, OCDM, and HRCSS.
- **[OFDM-OTFS]**
  - Codes that demonstrate the differences between OFDM and OTFS-based acoustic communication.

- **[tdoa-localization]**
  - Codes for a device-based two-way-ranging localization solution, including C codes for raspberry pi-based anchors (in subfolder **[anchor]**), target Android devices (in subfolder **[android_app]**), and JAVA-based windows server (in subfolder **[server]**).
- **[Two-way-ranging Android]**
  - A two-way acoustic ranging system in Android.
- **[phase_modulation.m]**
  - A code example for ZC-based modulation.
- **[waveform_example.m]**
  - Codes that generate typical spread spectrum modulated waveforms.

Index

R. Zheng, C. Cai, *Acoustic Sensing on Commodity Devices and its Applications*, Wireless Networks, https://doi.org/10.1007/978-3-031-96875-4

9783031968747